T0147106

THE HUNT FOR THE GOLDEN MOLE

The Hunt for
the Golden Mole

*All creatures great and small, and
why they matter*

Richard Girling

COUNTERPOINT
BERKELEY

Library of Congress Cataloging-in-Publication Data

Girling, Richard, author.
The hunt for the golden mole : all creatures great and small
and why they matter / Richard Girling.
pages cm
Includes bibliographical references and index.
ISBN 978-1-61902-450-2 (alk. paper)
1. Golden moles—Somalia. 2. Extinct animals—Somalia. 3. Hunting—
Moral and ethical aspects. 4. Nature conservation—South Africa.
5. Biodiversity—South Africa.
I. Title.
QL737.A352G57 2014
591.68--dc23
2014022506

Paperback ISBN: 978-1-61902-585-1

Typeset by Palimpsest Book Production Ltd, Falkirk, Stirlingshire

COUNTERPOINT
2560 Ninth Street, Suite 318
Berkeley, CA 94710
www.counterpointpress.com

Printed in the United States of America

For Caroline

CONTENTS

CHAPTER ONE

Dr Storer in the Chair

Hanno the Navigator, the Carthaginian explorer of the fifth century BC, also known as King Hanno II, is distinguished by having a crater on the moon named after him. He earned that honour by a feat of seamanship which, 2,500 years ago, established him in the line of pioneers that would lead via Leif Eriksson, Marco Polo, Vasco da Gama, Magellan, Cook, Livingstone, Scott and Amundsen all the way to Armstrong and Aldrin. Like the astronauts, Hanno could report a giant leap for mankind, but, unlike Armstrong's short stride into a cosmic future, his giant leap was backwards.

History over this number of centuries cannot be viewed in high definition. Fragmentary evidence and the agglomeration of myth leave only faint if highly coloured outlines, like lipstick on the rim of time. But what seems likely is that somewhere between 500 and 480 BC, Hanno set out from the Mediterranean with a fleet of sixty ships which, having passed through the Straits of Gibraltar, made a southward turn into the Atlantic. His mission was to explore and possess as much of north-west Africa as he could lay his hands on. Opinions vary on how far south he got but, in the light of events 2,300 years later, it seems likely that he reached the southern margins of what is now the Gulf of Guinea, and was off the coast of Gabon when he turned for home.

It was somewhere near here that the sailors found an island occupied and fiercely defended by a race of unusually hairy and furiously savage men and women. By Hanno's own account, the men were heavily outnumbered by the women and possessed of a ferocity that surpassed all reason. 'We pursued but could take none of the males; they all escaped to the top of precipices, which they mounted with ease, and threw down stones.' The Navigator was not inclined to sentimentality or squeamishness. 'We took three of the females, but they made such violent struggles, biting and tearing their captors, that we killed them, and stripped off the skins, which we carried to Carthage: being out of provisions we could go no further.'

History now fast-forwards to 1847. An American Protestant missionary, Thomas Staughton Savage, is busy bringing the Bible to West Africa when he meets a group of tribespeople worshipping the upper part of a skull mounted on a pole. But there is more to Savage than holy zeal. Like many churchmen in the nineteenth century, he is an avid amateur naturalist with a devout interest in all Creation. The skull fragment intrigues him. 'With considerable trouble', according to one contemporary, he manages to take possession of it, and the saucer of bone is soon making its way across the ocean to Massachusetts. With it go some other miscellaneous fragments and a paper written by Savage, which will be read on his behalf to the Boston Society of Natural History.

The minutes of the meeting held on 18 August 1847 ('Dr Storer, Vice President, in the Chair') record that the discovery was made 'in Empongwe, near the river Gaboon, Africa'. Laid out before the learned members are 'four Crania (two male and two female) . . . also the long bones of the extremities, a male and female pelvis, and some other bones'.

Unlike Hanno, the gentlemen are closely familiar with the races of great ape known collectively as Orangs. They know a chimpanzee when they see one, but they have never reckoned on anything as big as this. The account continues: 'This animal is known to the natives under the name of Ẽngeena, and is much larger and more ferocious than the Chimpanzée. Its height is above five feet, but it is remarkable for the disproportionate breadth of the shoulders, which is double that of the Chimpanzée. The hair is coarse, and black, except in old individuals, when it becomes gray. The head is longer than that of an ordinary man by two inches, and is remarkable for having a crest of coarse hair over the sagittal suture, which meets at right angles a second, extending over the upper part of the occiput, from one ear to the other. The fore-arm is much shorter than the arm, the hand is remarkable for its great size, and the thumbs larger than the fingers. A slight tuft of hair exists at the extremity of the *os coccygis* – no tail, no callosities. Its gait is awkward and shuffling, supporting itself on the feet and fingers, and palms of the hands; but not, like the Chimpanzée, resting on the knuckles.'

And then we hear the echo, rolling back across the millennia. 'They live in herds, the females exceeding the males in number. Their habitations, like those of the Chimpanzée, consist of a few sticks and leafy branches, supported by the crotches and limbs of the trees, which afford no shelter, and are occupied only at night. They are exceedingly ferocious, and objects of terror to the natives, who seldom encounter them except on the defensive. The killing of an Ẽngeena is considered an act of great skill and courage, and brings to the victor signal honor.'

The echo rolls on. 'The Orangs are regarded by the natives as degenerated human beings. The Encheeco, or Chimpanzée, being less ferocious, and more intelligent, is supposed to have the spirit of a *Coast-man*, but the Ẽngeena that of a *Bush-man*. Their flesh, when obtained, is eaten by the natives, as well as that of the Chimpanzée.'

And thus it happened, less than 170 years ago, that science 'discovered' the greatest of the great apes, surely the very same species of wordless, hirsute militant that Hanno's interpreters had called *Gorillae*. It was a classic example of a then typical event – another species new to science – and typically sent out a worldwide pulse of excitement. Notwithstanding Savage's fondness for the psalms, natural science was the rock and roll of the age. Monsters and curiosities in menageries and museums were irresistible crowd-pullers, bigger even than music-hall stars or charismatic preachers.

For a natural historian it was heaven indeed to be alive. Never had there been such an appetite for discovery, enlightenment and creative chaos. Even biblical literalists understood the earth to be, in their terms, a thing of great antiquity, yet its depths and extremities remained as mysterious as the moon. To find a new species in many parts of the world, all you had to do was walk outside and look. Men like Hans Sloane, Charles Darwin, Alfred Russel Wallace, Henry Walter Bates and Joseph Banks did a great deal more than just look. They *observed*. And, having observed, they noted, illustrated, collected and catalogued. It was a passion that seized some of the not-so-great minds, too. By sea and land, from Britain, Europe and America, adventurers poured into the unmapped forests, savannahs and wetlands of Africa, Asia, South America and Australasia. Some were men of science, some were men of commerce, some were rascals.

The decks and holds of ships were packed with animals alive, dying, dead or dismembered. In London, Abraham Dee Bartlett, the extravagant character about to begin a forty-year career as superintendent of London Zoo, received his first gorilla in1858. It reached England as heroes tended to do, like Nelson and Byron in a barrel of spirits. A photograph in one of Bartlett's books shows the author easing the animal from its cask. The strangely hairless gorilla is posed with one hand raised, apparently gripping

Abraham Dee Bartlett, superintendent of London Zoo, receives his
pickled gorilla in 1858

what looks like a pitchfork handle, as if trying to pull itself upright, while the other hand 'holds' the lid of the barrel. Bartlett was by profession a taxidermist, ever alert to the importance of presentation.

Necessarily, he was also alert to the tricks of his trade. When Richard Owen, superintendent of the natural history department at the British Museum and inventor of the word 'dinosaur', introduced him to a 'Monsieur du Chaillu', who was 'desirous to have his Gorilla skin properly stuffed', Bartlett caught the reek of vaudeville.

'I called M. du Chaillu's attention to the face of the animal, which I told him was not in perfect condition, having lost a great part of the epidermis. In reply he, M. du Chaillu, assured me that it was quite perfect, remarking, at the same time, that the epidermis on the face was quite black, and that the fact of the skin being black was a proof of its perfectness.

'I, however, then and there convinced him that the blackness of the face was due to its having been painted black; finding I had detected what had been done, he at once admitted that he did paint it at the time he exhibited it in New York.'

On another occasion Bartlett agreed to buy some fowls from a Japanese dealer, but only on condition that he was first allowed to dip their improbable six-metre-long tails in water heated to the melting point of glue. Dealer and fowls immediately took wing. Bartlett was not alone among experts in developing habits of caution. Where there is wonder, there is also disbelief. Many accounts of outlandish creatures recorded in the furthest corners of the world were received, at best, with scepticism. This was true even when there was a specimen to show. Fairground freaks had taught people not always to trust the evidence of their own eyes. One of the most notorious frauds was the 'Feejee Mermaid', which tripled the takings at Phineas T. Barnum's American

Museum in New York in 1842. Like many other exotic creatures, this one had been assembled by a Japanese fisherman – a monkey's body, finely stitched to a fish's tail. Its dried-up, withered appearance and repellent ugliness did nothing to deter the crowds that queued around the block to see it. As more and more bizarre specimens were uncrated, scientific wariness and fear of hoax made stubborn obstacles to credence.

Who, for example, would believe a beaver with a duck-bill stuck on to it? Even in 1798 this had seemed a bit too rich to stomach, never mind that the man who sent the first platypus back to Britain – Captain John Hunter, Governor of New South Wales – was an unlikely hoaxer. The eminent naturalist George Shaw, Fellow of the Royal Society, co-founder of the Linnean Society and a future Keeper of Natural History at the British Muscum, who is credited with the first scientific description of the species, snipped at the pelt in search of stitches but still admitted he could not be certain of its authenticity.

A century later it was the turn of the okapi to confound the doubters. You can see why. A chestnut-coloured horse with the legs and rump of a zebra, living hitherto unseen in the high-canopy forests of Central Africa? Who would have believed it? A beautiful and exactly detailed painting of the animal, sent to London by the explorer-naturalist Sir Harry Johnston, met with derision and was denounced by the director of the Natural History Museum, Professor Ray Lankester, as a hoax. Only in 1901, when skin and skulls were presented to a crowded meeting of the Zoological Society of London, did the okapi, and Johnston, get their due. The explorer's reward was to have the species named in his honour, *Okapia johnstoni*.

Disbelief came easiest to those whose experience of natural history was limited to periodicals and visits to menageries and museums. The okapi, like the gorilla, was a large and conspicuous item, impossible to overlook. How could it happen that no one had spotted one before? It fell to the great geographer and naturalist Alfred Russel Wallace to try to explain. In an essay published in 1878, he helped his readers understand what a tropical forest was actually like.

The observer new to the scene would perhaps be first struck by the varied yet symmetrical trunks, which rise up with perfect straightness to a great height without a branch, and which, being placed at a considerable average distance apart, give an impression similar to that produced by the columns of some enormous building. Overhead, at a height, perhaps, of a hundred feet, is an almost unbroken canopy of foliage formed by the meeting together of these great trees and their interlacing branches; and this canopy is usually so dense that but an indistinct glimmer of the sky is to be seen, and even the intense tropical sunlight only penetrates to the ground subdued and broken up into scattered fragments. There is a weird gloom and a solemn silence, which combine to produce a sense of the vast – the primeval – almost of the infinite. It is a world in which man seems an intruder, and where he feels overwhelmed by the contemplation of the ever-acting forces, which, from the simple elements of the atmosphere, build up the great mass of vegetation which overshadows, and almost seems to oppress the earth.

So, would not such a paradise be alive with animals? What should any explorer need more than a pair of eyes and time to record what he sees? Where is the scope for mystery? Wallace's answer to this is of profound importance in the light of all that will follow. 'The attempt to give some account of the general aspects of animal life in the equatorial zone,' he says, 'presents far greater difficulties than in the case of plants. On the one hand, animals rarely play any important part in scenery, and their entire absence may pass quite unnoticed . . . Beast, bird, and insect alike require looking for, and it very often happens that we look for them in vain.' It is an observation that could be as easily applied to an English woodland or North American forest as to, say, the Amazon Valley. Here in the 1850s Wallace's friend and collaborator Henry Walter Bates had to cope with disappointment 'in not meeting with any of the larger animals of the forest. There was no tumultuous movement or sound of life. We did not see or hear monkeys, and no tapir or jaguar crossed our path.'

If Bates didn't clock any large animals, it's pretty certain he wouldn't have seen too many small ones either. 'There is in fact,' as he later acknowledged, 'a great variety of mammals, birds and reptiles, but they are widely scattered and all excessively shy of man.' As Wallace describes it, the elusiveness of an animal seems to increase in proportion to one's desire to see it. 'The highest class of animals, Mammalia, although sufficiently abundant in all equatorial lands, are those which are least seen by the traveller.' This simple truism, self-evident to any child who has gone in search of a rabbit, still lays a curse on scientists wrestling with ideas of survival and extinction. It also explains why so many of the earliest voyages of discovery were focused on birds and plants rather than animals.

Not all of Wallace's encounters with mammals were born of his own curiosity. Having been bitten on the toe by a vampire bat (the toe 'was found bleeding in the morning from a small round hole from which the flow of blood was not easily stopped'), he took to sleeping with his feet wrapped up. But there are times when human intelligence – even intelligence on the Olympian scale of Wallace's – is confounded by primitive animal instinct. Next time, the vampire bit him on the nose.

Even Wallace, however, could not make accurate observations while asleep, and his account of the vampires' behaviour might have been lifted from a Gothic novel. 'The motion of the wings fans the sleeper into a deeper slumber, and renders him insensible to the gentle abrasions of the skin either by teeth or tongue. This ultimately forms a minute hole, the blood flowing from which is sucked or lapped up by the hovering vampire.' In fact, as we now know, the animal lands and approaches its victim on the ground.

Despite all the handicaps – shy, reclusive and nocturnal species, the impenetrableness of thorny, steep, over-heated and unmapped terrain – Wallace in the latter half of the nineteenth century builds a picture of richness, variety and almost comical oddness. In tropical and southern Africa alone, he writes, 'we find a number of very peculiar forms of mammalia. Such are the golden moles, the Potamogale, and the elephant-shrews among Insectivora; the hippopotami and the giraffes among Ungulata; the hyaena-like Proteles (Aard-wolf), and Lycaon (hyaena-dog), among Carnivora; and the Aard-varks (Orycteropus) among Edentata.'

Slowly, species by species, zoology was emerging as a scientific pursuit fit for the attention of serious minds. In the space of five years in the 1840s, the number of dead mammals acquired by the British Museum increased from around a hundred a year to more than a thousand. Natural history occupied a third of the museum's entire floor space, and attracted as many visitors as all the other galleries put together. In its early years, the museum had erected lofty bureaucratic barriers against casual visitors – tickets had to be booked in advance by personal representation, and were granted in scarcely greater number than audiences with the Pope. Now all that changed. As John Thackray, late archivist of the Natural History Museum, would write: 'The authorities accepted that the museum had a twin purpose: instruction for serious academic people, and rational amusement for the masses. It was felt that exposing the middle and working classes to a comprehensive display of the works of creation might improve their moral fibre and, also, make them proud to be British.'

The works of creation. One is surprised only by Thackray's omission of capital letters. The scientific world was drifting into two opposing camps – those who believed that Nature was ordained and delivered by God, and those under the influence of Alfred Russel Wallace and Charles Darwin, whose theories of evolution were putting the Book of Genesis under sudden and shocking pressure. This was no storm in a teacup. The origin of species was – as it remains – fundamental to the way we think about our rights and responsibilities. Even Christian fundamentalists had to think again about the size of the Ark. It was seldom forgotten that God had granted to man 'dominion over the fish of the sea, and over the fowl of the air, and every living thing that moveth over the earth' (Genesis I:28). Only now was the true scale of that dominion becoming apparent.

For the time being, despite the eruption on to the public consciousness of Charles Darwin (the entire first edition of *On the Origin of Species* in 1859 sold out immediately), it was the Old Testament that kept its nose in front. In January 1860 the decision was taken to hive off natural history from the rest of the British Museum – thus, in the words of Thackray, 'separating the works of Man (books, manuscripts and antiquities) from the works of God (natural history)'. When the new Natural History Museum in Kensington eventually opened its doors in 1881, visitors found that the superintendent, Professor Owen, had taken this sacred duty all too literally. His museum was a sermon encased in glass, a holy diorama of miraculous Creation in which the scientific voice was mute.

All I have in common with men like Alfred Russel Wallace is that I like to watch birds and animals. I do so very often without really knowing what I am looking at, or understanding the behaviour of the creatures I'm spying on. Sometimes I regret it, but more often I cherish my own naivety. It preserves my child's eye, a kind of pickled innocence that keeps nostalgia at bay. There are always questions to be asked.

Even in childhood I knew I wanted to write. Apart from running and jumping, it was the only thing I was any good at. But it was like having an instrument with no tune to play. What was I going to write about? People talk too glibly of Eureka moments, flashes of inspiration, epiphanies, and I hesitate to lay claim to one. But neither can I deny what happened. The occasion was the Easter holiday of 1961. I was fifteen-and-a-half, on a camping holiday with three friends in the county of Devon, halfway down the toe of the English south-west. Our plan, hatched over borrowed Ordnance Survey sheets, was to explore the great granite wilderness of Dartmoor. As I would discover in later life, its jagged tors – skeletal outbreaks of rock poking from the hilltops like springs through a worn-out mattress – were not much to set against, for example, the man-eating cols of the high Alps. But to a boy raised in the lawn-and-borders gentility of suburban Hertfordshire, the exposure to southern England's last untamed wilderness was life-changing. Dartmoor then had only recently been designated a National Park, and the untracked plateau was still a place of high drama and deep, unsettling mystery. My first sight of it, as I trekked up a lane towards Hay Tor, was one I shall never forget. A trick of topography, coupled with an over-active imagination, made it appear that the huge rock itself was rising up out of the ground in front of me. Before this, I had never known any feeling for landscape, history or 'the environment'. That all changed in an instant.

It was as if I had woken up in a different life. Suddenly, unexpectedly, I found myself transfixed by the authentic voices of the living and the dead. Sitting with illicit, under-age pints in village pubs, I listened to the stories of men who had worked the moor all their lives – shepherds, cowmen, horsemen, builders of stone walls, layers of hedges, makers of cider, men whose knowledge and craft linked them in a chain of ancestry that stretched back over millennia. High on the moor itself, the exposed relics of earlier civilisations raised questions about what those earlier centuries had been like. I bought a pair of binoculars and went nowhere without them. Into my rucksack went a back-breaking library of field guides – birds, mammals, insects, wild flowers, trees. And yet despite the best efforts of my mother, who was a devil for looking things up, I never quite caught the habit of naming things. Small brown bird. Tall yellow flower. They were tiny brushstrokes on a huge canvas, and it was the canvas that interested me.

Wherever I travelled, instinct always made me step back, viewing from a distance rather than homing in on the detail. The field guides went back on to the shelf and rarely came down again. Time moved on and the interest deepened into love, and love into a mounting anxiety – an anxiety shared, I quickly realised, by many others – that the canvas was becoming patched and stained. No matter where you looked in rural Britain, holes were appearing in a picture that was tending increasingly towards the monochrome. I might not remember what the tall yellow flower was called, but I noticed quickly enough when it was no longer there. And so I began to write, and proceeded to a mostly enjoyable, though frequently frustrating, career as an environmental author and journalist. On my desk sits the same heavy pair of Russian-made binoculars that magnified the lost countryside of my youth.

I have switched sides now, to the extreme east of England in the county of Norfolk. Somewhere outside my window, in woodland, field and hedge, lurk all the grazers and small-fry of Britain's diminished fauna. There are deer – now becoming a national scourge because they have no predators beyond motor cars or men with guns. There are rabbits, hares, squirrels, hedgehogs, the occasional fox (heard more often than seen), and the grass is creased by moles. Sometimes there is a rat; sometimes the sudden dash of a stoat or weasel. Everything else – the scurrying tribes of mice, shrews and voles – remains invisible to all but cats, owls and kestrels. Less than 16 miles from here, at West Runton, twenty years ago in a sea-cliff, was found an 85 per cent complete skeleton of what in life, 600,000 years ago, had been a ten-tonne steppe mammoth, *Mammuthus trogontherii*, twice the weight of a modern African elephant. It makes me think about the endless churn of life; the comings and goings of species that live out their span and disappear. Somewhere in the future, the pestilential rabbit and grey squirrel will go the way of the mammoth and the sabre-toothed tiger, and some other opportunistic invaders, drawn north by a warming climate, will inherit their niche. But then came something more thought-provoking still. In the autumn of 2010, news arrived of a completely new, previously unheard-of fish-eating mammal found living in Madagascar. With an exquisite sense of timing, the announcement came just a few days after the Royal Society had published a paper from the University of Queensland, proposing that a third of all supposedly 'extinct' mammals were actually still alive.

At almost exactly the same time, a UN biodiversity conference in Nagoya, Japan, was earning some very different headlines. A fifth of all the world's vertebrates – mammals, birds, reptiles, amphibians, fish – were facing extinction. In forty years, world populations of vertebrates had shrunk by 30 per cent. Land mammals were down by a quarter, sea fish by a fifth, freshwater fish by two thirds. On average, fifty species of mammal, bird and amphibian were edging closer to the brink every year. The big picture no longer made sense. New species? Reborn species? Extinct species? All swimming in the same ecological broth, but in different directions? The headlines came and went, but the questions stuck in my mind. How could creatures returning from the dead be reconciled with the threat of mass extinction? Why do estimates of the total number of species vary so widely? How could we be certain that any of them have died out? I remembered the words of Alfred Russel Wallace. *Animals rarely play any important part in scenery, and their entire absence may pass quite unnoticed.* Soon an even trickier question presented itself. How could we be sure that an extinct species had existed in the first place? Where did Nagoya get its numbers from? Who calculated them? Were they accurate? Did it matter?

Well, of course it matters. Whether it's Genesis or genetics that underpins our thinking, whether it comes with a capital N or a small one, nature is scorched into our subconscious, an ineradicable component of our genetic inheritance. It's how we, and the societies we've created, have evolved. Through the 'dominion' we have either accepted as a gift from God or claimed through right of arms, we have negotiated our own survival. We have made mistakes, hunted to near extinction the very species

– North American bison, the great whales – that we have depended upon. But it pains us. Even without understanding how ecosystems work, we know it's wrong – *absolutely* wrong, in a sense deeper even than the moral codes of law and religion. It's why we swerve to avoid a pigeon in the road. Yes, for all kinds of reasons, it matters. Time and again I return to the figures. Nothing adds up, and I realise that the big picture is no longer enough. Suddenly I have an appetite for detail.

My early attempts to satisfy it provoke good-natured complaint from the postman, bent under the weight of books. The standard taxonomic and geographical reference, *Mammal Species of the World*, comes in a hefty two-volume box-set from Johns Hopkins University. Nervously I flick the pages of Volume One, wondering how and where to begin. Turning to the very first entry on page one, I find:

ORDER MONOTREMATA Bonaparte, 1837
COMMENTS: Reviewed by Griffiths (1978). The order is the sole extant representative of the Subclass Prototheria (all other living mammals belong to the subclass Theria). McKenna and Bell (1979) divided the order into two (Platypoda and Tachyglossa); the date of divergence of the two living families is unknown, and conservatively they are retained here in a single order.

Reading on, I realise we're talking about echidnas and platy-puses, but I realise also, from the profound depths of my ignorance, that scientific detail is going to be hard on the diges-tion. Then – glory! – I remember Alfred Russel Wallace, for whom much of the world truly was a blank canvas, and I return to him as to a kind of intellectual comfort blanket. Let me begin where he began, and be led from there through zoological history.

What were the 'very peculiar forms of mammalia' that struck him so forcefully in tropical and southern Africa? 'Such are the golden moles, the Potamogale, and the elephant-shrews . . .'

Golden moles. I turn back to *Mammal Species of the World,* and claw my way to page 77:

SUBORDER CHRYSOCHLORIDEA Broom, 1915
COMMENTS: MacPhee and Novacek (1993) erected the suborder Chrysochloromorpha for golden moles, but following Simpson's (1945: 32–33) nomenclatural principles for categories above superfamilies, Chrysochloridea is the senior synonym.

Chrysochloridea it is, then. There are a good few of them – many more, I suspect, than even Wallace would have imagined. Most, but not all, have common as well as Latin names, usually in honour of their discoverers, territories or physical peculiarities. First up is Arend's golden mole, then Duthie's, Sclater's, Cape, Stuhlmann's, Visagie's, Giant, Rough-haired, De Winton's, Van Zyl's, Grant's, Fynbos, Hottentot, Marley's, Robust, Highveld, Congo, Yellow, Somali, Gunning's and Juliana's. Their territories range all the way down from the Gulf of Guinea, scene of Hanno's first brush with the *Gorillae*, through equatorial and sub-equatorial Africa to the Cape. But there are two exceptions, which, weirdly, appear to have no ranges at all.

Visagie's golden mole (*Chrysochloris visagiei*) – 'known only from the holotype'.

Somali golden mole (*Calcochloris tytonis*) – 'known only from the type specimen'.

By now I know that 'holotype' and 'type specimen' are the same thing. In each case they mean the original collected example from which the species was first described and introduced to science. What we are being told is that, throughout the whole of the scientific age, Visagie's and the Somali golden moles have each been seen only once. *One animal* constitutes the entire species. Conservatively, their status is recorded as 'critically endangered'. I will discover later that, though this degree of rarity is not a common phenomenon, it is not a rare one either. An astonishing number of species are accorded their identity on astonishingly sparse scraps of evidence. I turn next to the world authority on extinction and survival, the IUCN (International Union for Conservation of Nature) *Red List of Threatened Species*.

It confirms that Visagie's golden mole is known from a single specimen collected from the Northern Cape and described in 1950. We do not learn whether the animal was alive or dead, or even complete in all its parts, but a few drops of scepticism leak through the author's dry academic prose. 'Several field trips to ground-truth the occurrence of this species have yielded no specimens, or even signs of golden moles, suggesting either an error in recording provenance, or that the original specimen was transported there by anthropogenic means or even perhaps floodwaters of the Renoster River . . .'

If that is peculiar enough, then it's nothing to compare with its Somali cousin. Again the *Red List* confirms the uniqueness of the specimen, found at Giohar, Somalia, in 1964. But this time it adds an intriguing – not long ago I would have said *unbelievable* – detail. Under 'taxonomic notes', it remarks that the Somali golden mole, *Calcochloris tytonis*, is 'known only from a partially complete specimen in an owl-pellet'.

And that's it. Not only has no one ever seen a live example, no one has even seen a whole dead one. All that exists is some crumpled fragments coughed up by an owl. But exists where? It comes back to me in sleepless nights. First I am interested, then fascinated, then obsessed. Somewhere in a drawer, in a museum somewhere in the world, the owl pellet must be kept. *And I want to see it.* My naivety at that stage was still intact, so I thought it would be easy. I called the IUCN to ask where the specimen might be found. They didn't know. Was there not some compendious work of reference that listed all holotypes and their locations? There was not. Next I tried the Natural History Museum, then the Zoological Society of London. No one knew.

So here began both a mystery and a quest. There were several reasons why I resolved to try to find the Somali golden mole. There was the sheer exhilaration of the chase, the unravelling of a mystery, the bizarre improbability of a species catalogued from such minimal remains. But there was something deeper, something not quite thought through but naggingly insistent. At a time when one species, my own, was being forced to reconsider its relationship with every other, what was the moral of the story? How could I answer the question, put to me with some belligerence by a neighbour at a dinner party: *Why should I care about a species so obscure that no one has ever seen one?* Why do we need spiny mice, bearded pigs, groove-toothed trumpet-eared bats, glacier rats, or any of the dozens of other mammals that the IUCN tells us are on the downward slope?

Already I had half an answer, but I wanted to find a whole one.

CHAPTER TWO

Rhinoceros Pie

Sir Stamford Raffles, the founder of Singapore and discoverer of the clouded leopard, was the unstoppable force behind the establishment of the Zoological Society of London in 1826. He lived only long enough to chair its first two meetings before a stroke – 'apoplexy' in the language of the time – killed him on the eve of his forty-fifth birthday. But he had taken the crucial first step. Sir Humphry Davy and the Marquis of Lansdowne continued what he had begun, and the world's first scientific zoo opened at Regent's Park in 1828. Initially, the word 'scientific' was rigidly interpreted. Only fellows of the society were permitted to enter – a situation that would last until 1847. Even then, visitors needed a letter of recommendation and were barred on Sundays. It was undemocratic, and the science was rough round the edges, but it was progress. People began to think more carefully about animals – their physiology, their self-awareness, their behaviour – and zookeepers set out on the rocky road to enlightenment. It was an example that soon would be followed in other new zoos throughout Europe and America.

On a warm August day 163 years later, the Broadwalk in Regent's Park is a dawdling caravan of parents and children, all heading towards the zoo. Those bored or exhausted by the long trek from the bus or underground are kept moving by a promise which in all the years has never lost its potency. *Shall we go and see the gorillas?* I hear it time and again. The children will be disappointed only by the inert disinterest of the animals on the other side of the glass. My own hope – to see a living example of one of the surviving species of golden mole – has already been dashed. The zoo has told me it doesn't have one. And it gets worse. According to the online International Species Information System (ISIS), neither does any other zoo in the world. Golden moles may be 'vulnerable', 'endangered', or 'critically endangered', according to IUCN conservation criteria, but I can detect no effort to conserve them.

I don't do much better with the 'peculiarities' that so diverted Alfred Russel Wallace in southern Africa. Where the aardvarks ought to be, I see only meerkats. There are no hyenas, aardwolves or elephant shrews, though for compensation there is a magnificent okapi – a species known to Wallace only in the last few years of his life.

London Zoo now would astonish its nineteenth-century superintendent Abraham Dee Bartlett. Few of the original buildings survive, and many of the stars of the early collection – bears, elephants, hippos, rhinos, pandas – have been taken away. Some, like the quagga, are globally extinct. For pioneers such as Bartlett, keeping animals was a process of trial and error. His exhibits were not captive-bred specimens of known provenance, well-documented health and studied habit. They were wild-caught strangers wreathed in mystery. Bartlett recorded the arrival on 22 May 1869 of the zoo's first panda. It was not in good shape.

'I found the animal in a very exhausted condition, not able to stand, and so weak that it could with difficulty crawl from one end of its long cage to the other. It was suffering from frequent discharges of frothy, slimy faecal matter. This filth had so completely covered and matted its fur that its appearance and smell was most offensive.' He identifies the species as *Ailurus fulgens*, the small, teddy-bear-like red panda, not the giant panda *Ailuropoda melanoleuca*, but most people today would be able to guess what it ate – mostly bamboo, supplemented by eggs, birds and small mammals. Bartlett, however, knew none of this. 'The instructions I received with reference to its food were that it should have about a quart of milk per day, with a little boiled rice and grass. It was evident that this food, the change of climate, the sea voyage, or the treatment on board ship had reduced the poor beast to this pitiable condition.' With no textbook to consult, Bartlett could only guess what to feed it with. He went to work with a zeal that might have earned the envy of his contemporary, Isabella Beeton. First he tried raw and boiled chicken, rabbit and 'other animal substances', but the panda would have none of them. 'I found, however, it would take arrowroot, with the yelks [sic] of eggs and sugar mixed with boiled milk; and in a few days I saw some improvement in its condition. I then gave it strong beef-tea well sweetened, adding pea-flour, Indian-corn flour, and other farinaceous food, varying the mixture daily.'

Soon the panda was well enough to be let out into the gardens, where it straightaway attacked the fruit and foliage. It liked particularly the large yellow berries of a tree Bartlett named as *Pyrus vestita*, now better known as *Sorbus cuspidata*, a native of China, the country whose south-western provinces are the panda's home. 'He would grasp the bunch in his paw, holding it tightly, and bite off these berries one by one; so delighted with this food was he, that all other food was left as long as these berries lasted.' It enabled Bartlett to conclude 'that berries, fruit, and other vegetable substances constitute the food of this animal in a wild state'. For zookeepers of the nineteenth century, this was how it went. They would work like field naturalists on the basis of observation wherever that was possible, and by trial and error when it wasn't.

They also learned to respect wild animals' natures, and did not expect them to cosy up like family pets. Bartlett noted somewhat ruefully the panda's 'fierce and angry disposition', though he believed this to be a peculiarity of the individual and not necessarily typical of the species. Even an attack was the subject of careful study: 'When offended, it would rush at me and strike with both feet, not, like a cat, sideways or downwards, but forward, and the body raised like a bear, the claws protruding, but not hooked or brought down like the claws of a cat . . .'

One of Bartlett's many scientific acquaintances, and a frequent visitor to the zoo, was a naturalist called Frank Buckland, who (as we shall see) studied all things zoological with a passion that verged on mania. He was also a prolific writer who liked to publish his correspondence with other enthusiasts. One of these was Bartlett, who sent him a long description of how he had treated a hippopotamus with a broken tooth. Deciding that extraction was the only answer, and working from behind an oak fence, he had proceeded with 'a fearful struggle' involving an enormous pair of forceps more than two feet long. The operation began well. He quickly managed to get a grip on the fractured incisor, which he intended to remove 'with a firm and determined twist'. The hippo, alas, was both firmer and more determined, and the forceps were wrenched from Bartlett's grasp. It was a tribute to the quality of the carpenters' work that the fence stood up to the animal's charge and Bartlett survived to try again. This time he had a little more success – the tooth was actually loosened – but again the patient had the better of him and the forceps went flying. The third attempt artfully capitalised on the animal's rage. 'Looking as wild as a hippopotamus can look', the monster advanced upon Bartlett with its jaws at full stretch, wide enough to swallow a canoe. The 'coveted morsel', as Bartlett put it, was then easily grasped and, 'with a good sharp pull and a twist', drawn out. Like everything else about the animal, it was huge. 'One of the most remarkable things,' Bartlett wrote, 'appeared to me to be the enormous force of the air when blown from the dilated nostrils of this great beast while enraged. It came against me with a force that quite surprised me.'

One cannot quarrel with Buckland's opinion that the super-intendent 'deserves great credit for his ingenuity and the surgical skill he displayed with his huge patient'. In many ways Buckland himself was no less adventurous. It was his habit, for example, to cook and eat animals that had died in the zoo, and he once entertained an audience at Brighton by serving them rhinoceros pie. His real passion, however, was what he liked to call *hippophagotomy*, or the consumption of horseflesh. This had begun with an invitation to lunch, and a challenge, from Bartlett himself, who had placed 'two exceedingly fine hot steaks' on the table. One was 'rump-steak proper', and the other a slice of horse. In a blind tasting, both men preferred the horse. 'Uncommon good,' said Buckland.

But he knew his enthusiasm would not be widely shared. Indeed, it gave him the idea for a novel method of deterring crime. It was perfectly simple. The 'lower classes', he argued, had an irrational horror of eating horse, which they regarded as fit only for cats. Therefore all that was needed to curb their anti-social tendencies was to serve the stuff in prisons. 'Be assured these fellows who would garrote [sic] you, murder your wives and children, or commit the most fearful crimes, would shudder at the thought of dining upon horseflesh.' The theory was never put into practice, though its basic premiss was vividly demonstrated early in 2013 when some meat products in England and France were found to contain more horse than beef. In the face of public outrage, supermarkets competed to out-apologise each other and be first to clear 'value' burgers, lasagne and spaghetti bolognese from their shelves. Regulatory authorities across Europe swooped on shops, processors and abattoirs, and the British prime minister was urged by the leader of the opposition to 'get a grip'.

It took a lot to make Buckland himself shudder. Squeamishness was not in his vocabulary. When an old lion died at the zoo, he was present at the dissection to peel the skin from the foot and fiddle with the tendons (they worked 'with the ease of a greased rope in a well-worn pulley'). When lions broke out of their cage at Astley's Royal Amphitheatre in Westminster Bridge Road, Lambeth – birthplace of the circus ring – he was on hand to examine the corpse of the unlucky stable-hand who got in their way. 'It will probably interest the reader, to read some remarks on the nature of the wounds, and on the probable way, judging from these wounds, in which the lion seized the man.' And on he went, scratch by scratch, bite by bite. 'I account for there being so many more wounds on the left side than on the right side by assuming that the lion (as is its habit) cuffed him first on the right side and caught and held him on the left, just as we see a kitten playing with a ball of worsted.'

Buckland was a disputatious fellow, probably not the kind of man it would have been wise to accuse of hypocrisy to his face. A hundred and thirty years after his death, however, he seems fair game. Like most gentlemen of his time, he was an eager sportsman who enjoyed nothing better than a duck-shoot at dawn, proudly recording the rain of teal and widgeon from the sky. But if he was easy of conscience, he was markedly less forgiving of others. In the third volume of his *Curiosities of Natural History*, he seems suddenly overcome by loathing for his fellow hunters:

> In reading the accounts of the mighty elephant in the jungle of India, of the watching for the beasts of the forest drinking at midnight at the lone desert fountain in Central Africa, of the fierce gorilla in the dense forests of the tropics, or of wild ducks and swans on some lonely lake or swamp, I often come on the most exciting description of the discovery of these creatures, feeding quietly and undisturbed in their native homes. What a chance, what an opportunity of learning their habits, and their loves, and their wars! But – No; man thirsts for their blood. A few lines further down the page of the book we read the old story – I mentally hear the ring of the rifle or gun – and in an instant a beautiful scene of Nature is ruthlessly dissipated. The frightened creatures fly hither and thither; what was but just now all happiness and quiet, resolves itself into bloodshed, turmoil and misery . . . Let a knowledge of the habits of an animal or bird be of far greater value to the sportsman-naturalist than the possession of its bleeding carcase.

It takes him only another eight pages to revert to type, uncritically recording how his late friend Dr Genzick of Vienna had killed a hippopotamus.

> . . .The ball struck the hippopotamus full on the head, and he sank instantly to the bottom, where he kicked up such a turmoil that, as Genzick said, 'one would have thought there was a steam-engine gone mad at the bottom

of the river'. However, the doctor never found the hippopotamus, though he hunted everywhere for him, but the
next year he discovered his whitened bones upon a sand-
bank some distance from the place where he had shot him.
He knew it was the beast he had shot the year before, for
he recognised the bullet he found in his skull as his own
make.

This does not mark out either Buckland or Genzick as a
moral degenerate. Even down to their inconsistencies, they
conformed to the spirit of their age. Men with a true and affec-
tionate interest in animals were a willing party to what any
civilised person now would deplore as unspeakable cruelty. They
saw no contradiction in this, still less hypocrisy, but only the
hard demands of necessity. Even among their own kind, death
was a frequent visitor who – *pace* Stamford Raffles – often called
unannounced, and seldom with the clean finality of a bullet.
Sentimentality was for novelists. If zoos and museums wanted
animals, then someone would have to go and fetch them.
Milksops need not apply. For a European or an American, just
getting to Africa and surviving there would require both a rugged
body and uncommon strength of mind. Add confrontations with
snakes, insects and man-eating lions, and the necessary qualifi-
cations excluded all but the most determined of adventurers.
I've seen no evidence that Frank Buckland himself ever strayed
far beyond Paris, but this did not stop him from describing the
dangers of further continents. Hippos in particular seemed to
fascinate him, and he cites a 'Mr Petherick' (presumably the
Welsh mining engineer, explorer and collector John Petherick)
as his source for a vivid account of the risks to men in boats. A
hippo, he explains, will attack in one of two ways. In shallow
water it will bound up to the boat, then 'rise open-mouthed and

endeavour to carry off some one on board'. In deeper water it will drive at full speed underneath and use its head as a battering ram. Mr Petherick told him of boats being instantaneously sunk, and of a man being cut in two by the animal's teeth.

Hunters' yarns make fishermen's tales seem like essays in modesty. One of the most shameless exponents of the bragger's art was Roualeyn Gordon-Cumming (1820–1866), the self-glorifying Old-Etonian son of a Scottish baronet who was proud to be known as 'the lion hunter'. His own written account, *Five Years of a Hunter's Life in the Far Interior of South Africa*, published in 1850, contains such a sustained crescendo of bare-chested boasting that the feminist characterisation of masculinity as 'testosterone poisoning' no longer seems quite so unreasonable. How to remove a wounded hippo from a pool? Easy! Plunging into the water with a knife, Gordon-Cumming cuts a slit in the animal's hide like a belt-tab on a waistband. A thong threaded through the loop is then passed up to his men, who form up like a tug of war team and haul the beast ashore. Perhaps these are lines to be read between, but Gordon-Cumming cannot be dismissed out of hand as a fantasist. His vast collection of hunting trophies and stuffed animals – weighing in all 27.21 tonnes – caused a sensation at the Great Exhibition in 1851, and no less a figure than David Livingstone attested that his book conveyed a 'truthful idea' of the hunter's life. It remains a classic of sporting literature, which even in its time caused amazement. To the armchair readers of Victorian England it was a tale of heroism drenched in the spirit of Empire. To a modern reader it can seem simply outrageous. No pages ever dripped with more blood; no writer ever found more glory in the taking of life, or showed less remorse for the suffering he caused; no book ever spoke more definitively of attitudes that, in a century of blazing attrition, would bring nature to its knees.

Gordon-Cumming nevertheless was both brave and resourceful. He lurched about southern Africa in a wagon train that time and again would lose its wheels or bog itself down in desert or river, drawn by uncooperative oxen whose lives were sucked out of them by tsetse flies. He spent nights in bothies roofed with elephants' ears, while his animals and men fell prey to disease, buffaloes and lions. In the course of four expeditions he lost forty-five horses, seventy cattle and seventy dogs. All that remained of his best wagon-driver one morning was a leg bitten off below the knee, still wearing its shoe. Gordon-Cumming himself, though weakened by rheumatic fever and too much rhinoceros meat, never faltered in his appetite for sport or his determination to enjoy it. When the barrel of a favourite gun burst, burning his arm and causing temporary deafness, he 'mourned over it as David mourned for Absalom', but then simply switched to 'the double-barrelled Moore and Purdey rifles, carrying sixteen to the pound' and made bullets for them by melting down his snuffers, spoons, candlesticks, teapots and cups. His accounts were declared by some commentators to be 'romantic'. They are certainly graphic.

> . . . I was loading and firing as fast as I could, sometimes at the head and sometimes behind the shoulder, until my elephant's forequarters were a mass of gore, notwithstanding which he continued to hold stoutly on, leaving the grass and branches of the forest scarlet in his wake.
>
> Having fired thirty-five rounds with my two-grooved rifle, I opened fire upon him with the Dutch six-pounder; and when forty bullets had perforated his hide, he began for the first time to evince signs of a dilapidated constitution.

The tally of elephants rises to fifty, then to a hundred. Many of them die after prolonged 'fights' that go on for hours, with the quarry constantly harried by dogs and fired upon by Gordon-Cumming from his horse. One such bout goes on from half past eleven in the morning until after sundown, by which time the 'venerable monarch of the forest' has received fifty-seven bullets. The high price of ivory adds an economic impulse to the sport, but Gordon-Cumming doesn't stop at elephants. Antelopes, rhinoceroses, hippos, giraffes, lions, wildebeest, buffaloes, zebras, kudus, elands, wild boars and crocodiles are all 'bowled over' by his fire. In one pool alone he kills fifteen hippos. Bullets pierce shoulders, legs, flanks, necks, breasts, eyes, mouths and brains. He does not say that killing is better than sex, but it certainly beats anything else he can think of. Taking your pick of five old bull elephants, he finds, 'is so overpoweringly exciting that it almost takes a man's breath away'.

He is not blind to the animals' beauty, but it seems only to increase his pleasure in killing them. 'I was struck with admiration at the magnificence of the noble black buck, and I vowed in my heart to slay him . . .' After a wounded lion has crawled off and died, he regrets the inadequate power of words to convey his feelings. 'No description could give a correct idea of the surpassing beauty of this most majestic animal, as he lay still warm before me.' There is an aesthetic of death. As beauty enhances the thrill, so ugliness must diminish it. Afterwards, returning to camp, he spots and kills an 'extremely old' black rhinoceros, but can find little or nothing to commend it. 'His horns were quite worn down and amalgamated, resembling the stump of an old oak tree.'

The numbers of dead were prodigious. The 27 tonnes of exhibits that drew the crowds in 1851 were not the total weight of animals Gordon-Cumming had killed but just the heads, horns, tusks and skins that he stripped from the bodies. Apart from some meat and fat taken for food, all the rest was left where it fell. Where the specimens were not of exhibition quality, he might not take anything at all. The only benefits were to vultures and hyenas.

And not all the animals died – or at least they did not die quickly. Gordon-Cumming forever complains of wounded animals dragging themselves away, leaving only trails of bloody footprints into the bush. He is 'very much annoyed at wounding and losing in the last week no less than ten first-rate old bull elephants'. Another time he bags five 'first-rate hippopotami', but only at the cost of wounding three or four more. Casually he maims a white rhinoceros but is so preoccupied with the elephants that he does not follow up to kill it. Lions, buffaloes, crocodiles and antelopes are wounded too, but never does Gordon-Cumming's regret extend further than his loss of a trophy. It certainly does nothing to dampen his pleasure. The wounding of eight elephants is all part of 'the finest night's sport and the most wonderful that was ever enjoyed by man'.

Gordon-Cumming of course deserves to be judged by the standards of his own day rather than ours, and in his own day he was a hero. One of the antelopes he shot turned out to be a new species which still bears his name: *Tragelaphus scriptus roualeyni*, the Limpopo bushbuck. 'Conservation' in the mid nineteenth century was not an issue or, in its modern sense, even yet a word. The forests of Africa teemed with fur and the oceans teemed with fin. Men by their own efforts could no more deplete this almighty horde than they could fly to the moon or warp the climate. Such things were the province of God alone. Few people now would argue with Jeremy Bentham's caution on the nature and quality of non-human life. 'The question,' he said 'is not "Can they reason?" nor, "Can they talk?" but rather, "Can they suffer?"' Philosophers and psychologists still argue about the conscious lives of animals, their self-knowledge and 'intelligence', but none now doubts they can feel pain. Many believe they suffer emotionally too. But what is commonplace in the twenty-first century was not so clear in the nineteenth. It is true that Bentham, the founder of utilitarianism, had been dead for nearly twenty years before Gordon-Cumming published his book, but the question was still hanging. Could animals suffer? René Descartes had been dead for exactly two hundred years, but his ideas still cast a shadow.

Descartes, celebrated as the founder of modern philosophy, has a valid claim to be regarded as a genius, one of the greatest minds of the seventeenth century. Three hundred and fifty years after his death, his achievements still merited six whole pages of *Encylopaedia Britannica*. Very few people, before or since, could claim to be his intellectual superiors. And yet he chose to devote his huge mental power to a densely argued theory which demonstrated to the satisfaction of all the leading scholars of his day that animals had no conscious life. To have a conscious life you needed an immortal soul, and animals had no immortal souls. They believed nothing, desired nothing, felt nothing. They were like machines. If you applied a stimulus, out would come the matching response. If you shot one, or stuck a knife into it, the noise it made was purely mechanistic, not a cry of pain. Lacking consciousness, animals could not feel pain. Anyone who thought otherwise was guilty of anthropomorphism. This comforting misapprehension paved the way for what opponents of laboratory procedures on living animals still like to call 'vivi-section'. If an anatomist wanted to study the innards of a dog, he could simply nail it up by its paws and open it with a knife.

By the same logic, a hunter in search of specimens could blaze away with no apprehension of cruelty.

That was the theory. The reality, I suspect, was somewhat different. The romantic poets of the eighteenth and early nineteenth centuries, well before Roualeyn Gordon-Cumming dipped his pen, had urged their readers to show kindness to animals. As Coleridge writes in *The Rime of the Ancient Mariner* (1798):

> He prayeth well, who loveth well
> Both man and bird and beast.
> He prayeth best, who loveth best
> All things both great and small.

What would be the value of kindness and love to creatures unable to respond? Philosophers could argue, and scientists seek for proofs, but there was little doubt in the public mind that what *looked* like pain in animals *was* pain in animals, and that pain meant suffering. The passions that drove Anna Sewell to write *Black Beauty*, probably the most widely read plea for animal welfare ever published, were burning long before the book appeared in 1877. It would be easier to acquit Gordon-Cumming if we could be sure he held the Cartesian view and was of innocent mind. But his writings demonstrate unequivocally his awareness that animals could suffer, and that in some instances their suffering is deserved. This is made obvious when his wagon-driver is eaten by a lion. In pursuit of the killer, he writes: 'I wished I could take him alive and torture him, and, setting my teeth, I dashed my steed forward within thirty yards of him and shouted, "Your time is up, old fellow."'

What would be the point of torturing an animal if you thought it could feel no pain? That 'old fellow', too, is typical of Gordon-Cumming's tendency to anthropomorphise, or to describe animals in terms of their characters. He certainly understands that dogs can suffer. 'On proceeding to seek for Shepherd, the dog which the lion had knocked over in the chase, I found him with his back broken and his bowels protruding from a gash in the stomach; I was, therefore, obliged to end his misery with a ball.' Whatever his reason for not extending this sensitivity to elephant, hippo or lion, it cannot be that he believed them incapable of suffering.

It is obvious from the celebrity Gordon-Cumming enjoyed that no great opprobrium attached to his spree. But it ill behoves the twenty-first century to accuse the nineteenth of double standards – we have enough of our own. My favourite example of moral confusion is from the 1980s at the University of Tennessee. At that time it was home (as it probably still is) to some of the world's most privileged mice. Their accommodation was temperature- and humidity-controlled. Their bedding was fresh, their diet a masterclass in nutritional exactitude. But there was, inevitably, a price to be paid. The mice were purpose-bred for the university's laboratory, and their ultimate destiny was to die in the service of human health. To compensate for this sacrifice, and for as long as they lived, their comfort would be guaranteed. Their welfare was legally protected, and nothing could be done to them without the informed consent of the university's animal care committee. At the end, attended by their own dedicated vet, they would be wafted to the hereafter on an overdose of anaesthetic. Few humans would live and die as painlessly.

But these were not the only mice at the University of Tennessee. In secret places beneath floors and furniture, behind skirting boards, lived another quite separate population – genetically identical to the five-star specimens in the laboratory, even directly related to them, but socially a world apart. These mice were pests whose health and well-being were of no concern to the US Department of Agriculture or to the university's animal care committee. Their welfare was left to the caretakers, who trapped them on sheets of cardboard spread with glue. The irony of their sticky end was not just that it would have been indefensible if practised on their upstairs cousins. It was that the gluepot victims had once been five-star mice themselves. Their fatal error had been to escape, and not to understand the small print of human ethics.

But the moral maze doesn't end there. The university housed yet another group of mice, procured for the benefit of the zoology department's snakes. It was a core principle of the animal care committee that animals should be fed their natural diets – which, for the snakes, meant live mice. The ethical proviso was that this must be done for dietary reasons alone, not for the sake of an experiment. If a researcher decided to increase the value of his snake project by studying, say, the fear responses of the mice, then there would be a further, seismic upheaval in the ethical landscape. The mice themselves would become the subject of an experiment, and being fed to snakes undoubtedly would cause them to suffer. The animal care committee therefore would need to hear a very convincing explanation before allowing the observations to continue.

The story of the Tennessee mice was told by an American psychologist, Harold A. Herzog, in the *American Psychologist* magazine. He drew the obvious conclusion. The moral judgements that humans make about other species 'are neither logical nor consistent . . . The *roles* that animals play in our lives, and the *labels* we attach to them, deeply influence our sense of what is ethical.' In plainer language, we are prejudiced. Our attitudes to animals are determined by the labels we attach to them – *pet*, *food, pest, vermin*. Shuffle them around, and the result is almost viscerally disturbing. Pony-veal? Cat-traps? A dog-shoot? Moral duplicity is inescapable. A few months ago I paid a man to put ferrets down the rabbit holes in my garden, to flush out and snap the necks of the wild breeding stock. Later the same day, like a repentant Mr McGregor, I offered fresh carrots to my neighbours' pet rabbits. Afterwards, with a glass of good red burgundy in hand, I enjoyed a pie made from one of the ferreter's victims. There is no moral consistency in any of this; only a kind of self-interested pragmatism. At the very same time as I was developing my fascination with the Somali golden mole, I was laying traps for the all too common-or-garden local mole, *Talpa europaea*. The one is rare to the point of invisibility; the other abundant to the point of nuisance. Two similar species; two very different attitudes.

There is no reason to wonder, therefore, how it was that Roualeyn Gordon-Cumming made such distinctions between dog and wild beast. They were irrational, but they were not incomprehensible. And he was in good company. Back home in Europe and America, public interest in zoological exotica was such that showmen like P. T. Barnum and the Ringling Brothers could make fortunes from it. And science, too, had much to learn from the hunters' specimens. In its early years even the Zoological Society of London owned more dead animals than live ones, and museums of natural history throughout the world relied on the bullet to fill their display cases. With all due reverence, one recent writer describes the great Central and North Halls of the Natural History Museum in London as a 'Valhalla for British natural history'. Here stand memorials to the museum's first superintendent, Richard Owen, and to the secular gods Charles Darwin and Alfred Russel Wallace. Owen, who loathed Darwin and all his works, stands gowned on his pedestal, hands outstretched like a prophet in mid-sermon. Darwin himself sits cross-legged in his chair, hands in lap, as if resting from the burden of his own huge brain.

But there is another, more flamboyant figure, a man in a bush hat brandishing a rifle above a bas-relief of lions. This is the hunter, explorer and naturalist Frederick Courteney Selous (1851–1917), the real-life inspiration for Rider Haggard's fictional adventurer Allan Quatermain. The presence in Valhalla of a famous killer, arguably the deadliest white hunter ever to load a gun, is not an aberration. In honouring him with a bronze, the museum was simply acknowledging its debt. Men like Wallace and Darwin may have given the museum its *raison d'être*, but it was men like Selous who filled its display cabinets. His

Great white specimen hunter – bust of Frederick Courteney Selous at the Natural History Museum in London. It was his rifle that stocked the display cabinets

marksmanship in Africa provided the museum with jackals, hunting dogs, hyenas, lions, leopards, cheetahs, buffaloes, antelopes, gazelles, wildebeest, reedbucks, waterbucks, bushbucks, kudus, elands, elephants, giraffes, warthogs, hippos, zebras, rhinos and elephants. From elsewhere in the world came wolves, otters, lynxes, bison, goats, chamois, deer, moose and reindeer.

Alas for me, he never shot a Somali golden mole. More ominously in retrospect, neither he nor Gordon-Cumming ever killed a bluebuck (*Hippotragus leucophaeus*). This was not by accident or because they thought it to be not worth the price of a bullet. In 1799, twenty-one years before Gordon-Cumming's birth and fifty-two years before Selous's, this large South African antelope, also known as the blaubuck or blaauwbock, had passed from the veldt into the history books – the first large mammal in historic times to be hunted to extinction.

CHAPTER THREE

Beings Akin to Ourselves

Men like Gordon-Cumming and Selous must have known something about the way species interacted, and have had some idea that the tiniest scraps of life at the bottom of the food chain were in some way important to the behemoths at the top. But they were showmen as much as naturalists, and their audiences were not much attracted by small and drab. Who would queue to see a mole? Who knew or cared anything about 'ecology' (the word did not even exist until coined by the German biologist Ernst Haeckel in 1866). People wanted drama – living colour and bold brushstrokes. Everything else could bide its time. Most species of golden mole were not described until after Gordon-Cumming's death, and *Calcochloris tytonis* had to wait until 1968. It may be commonplace now to talk about 'biodiversity', 'ecosystems' and 'symbiosis', but these are modern concepts built jigsaw-fashion over decades. The hunters saw only what was in front of them, one species at a time, biggest first. Who could be surprised that they showed little interest in anything smaller than a dik-dik?

Given the challenges of surviving the African climate, never mind the impossibility of heavy haulage through unmapped forests and plains, it is easy to understand why most of the trade was in heads, horns, tusks and skins. A man with an ox-cart and a rifle – even one as resourceful as Gordon-Cumming or Selous – was not going to bring home a fully grown live hippopotamus. The Natural History Museum would have been nothing without marksmen and taxidermists. There is no irony, intentional or otherwise, in the elevation of Selous to its pantheon of heroes.

All the same, showmen knew very well that live action would sell better than static display, and zoos by definition needed life. There was a powerful incentive to 'Bring 'em back alive', as the twentieth-century Texan adventurer Frank Buck would put it in the title of his bestselling book. The problem was that wild animals did not travel well. What started out alive was more than likely to be delivered dead, and few of the survivors would last long in captivity. It was a vicious circle. The high mortality rate only increased the demand for replacements, thus inflating the prices and attracting the kind of entrepreneur for whom the scent of a fast buck was made no less sweet by the stench of corpses. But again we have to understand the spirit of the age. Men in the early nineteenth century did not inflict upon other species any cruelty they were not willing to inflict upon their own. The slave trade in the British Empire was abolished only in 1807, and slavery itself remained legal until 1833. Even during the lifetime of my grandfathers, in the 1890s, African and Arab slave traders were still resisting attempts to close them down. Despite the establishment of the Royal Society for the Prevention of Cruelty to Animals in 1824, and of the American Society for the Prevention of Cruelty to Animals in 1866, there were very few curbs, legal or moral, on the worldwide trade in birds and beasts.

The heightened sensitivity of the twenty-first century would have been as impossible for the nineteenth century to imagine as the loss of Victoria's empire. Our double standards might have struck them as absurd. On my way to see London Zoo's okapi, I linger in the giraffe house, and I ask myself: What do I think about this? Where is my moral centre of gravity? I try to work out how I score. On killing for sport I have a clean sheet. Never done it; never will. People who stand in fields and pick off tame pheasants strike me as, at best, laughable. Great white hunters, eh? Beyond that I am in difficulty. My misgivings about pet-keeping are compromised by the cats and guinea pigs beneath the plum tree in my garden. My tolerance of other species sharing my space is widely variable. I won't tread on an ant if I can avoid it, and I work at a desk under a tent of undisturbed spiders' webs, but I've killed rabbits and moles, and there are mouse-traps in the kitchen. I eat meat, lots of it, and I have made a public defence of animal laboratories. Nobody could mistake me for a Jain. The zoo therefore triggers a maelstrom of conflicts. The giraffe is not a threatened species. Its range in sub-Saharan Africa has been sadly reduced, and there may be uncertainties about its long-term future, but the IUCN calculates a viable wild population in the region of 100,000 and classifies it as a species of least concern. Its survival does not depend on conservation by zoos. I ought therefore to feel unease,

High risk – big animals were frequent victims of accidents at the docks

perhaps even indignation, at the sight of these miraculous crea-tures in confinement so far from their natural habitat. But I don't. Wherever in the world I go, I am unlikely to come across a happier contrast than between these sleek, apparently contented animals and their unfortunate historical forebears who might have died for nothing more than their fly-whisk tails. I pass on with contradictions unresolved. Situation normal.

All trafficked animals in the nineteenth century suffered in handling, but the giraffes' great height, their long necks and gangling limbs, made them particularly vulnerable. Being the most awkward of cargoes, they all too easily fell from dockside cranes. In 1866 two were killed in a fire at London Zoo. In 1876 at Hamburg, three more broke their necks against a wall. In the wild, where they loomed high above the low African skyline, they drew a dangerous amount of attention to them-selves. Along with the golden moles, aardvarks and hippopotami, they were counted among the 'very peculiar forms of mammalia' celebrated by Alfred Russel Wallace, and were conclusive evidence of Africa's weirdness. No animals ever caused more of a stir in Europe than the first giraffes, delivered in 1827 by the Viceroy of Egypt as gifts to the British and French govern-ments. They were also easy to shoot, and provided men like Roualeyn Gordon-Cumming during their travels with a regular supply of meat. He reminds us again of the cheapness of animal life:

> As we neared the water I detected a giraffe browsing within a quarter of a mile; this was well, for we required flesh . . . He proved to be a young bull, and led me a severe chase over very heavy ground. Towards the end I thought he was going to beat me, and I was about to pull up, when suddenly he lowered his tail, by which I knew that his race was run.

Urging my horse, I was soon alongside of him, and with three shots I ended his career.

Another day he chased and shot 'the finest bull' in the herd, but took nothing from it but the tail. It was against this ingrained tradition of kill-as-you-go that the live animal traders moved in and developed their businesses. There would be changes in practice but not in outlook. The merchants were not monsters – memoirs reveal some concern for animal welfare – but they were pragmatists who accepted the crude realities of their trade. Purposely or by accident, they would waste as many lives as it took to satisfy their customers.

Where big animals such as giraffe and elephant were concerned, the safest option for a trapper was to target the young. If anything, this actually increased the number of dead. Before you could catch a baby you had to kill its mother and the herd leaders that would defend it. Losses of breeding stock were immense. A zoo official in London reckoned that the price of one live orang-utan was four killed in the wild. Yet this was only the beginning. There is no record of how many captive animals died on the journey from forest or veldt to the coast, but on the evidence of later chroniclers such as Frank Buck we may assume the number was huge. Even that was not the worst of it. Of those that survived long enough to go aboard ship, half were lost at sea.

Good intentions were no guarantee of a humane outcome. In January 1867 Frank Buckland visited Charles Jamrach – a well-known and reputable London animal trader who counted many zoos among his clients. In Jamrach's shop Buckland noticed the skulls of two Indian rhinoceroses. How had his friend come by these? It was a terrible story, which began when Jamrach sent his son to India to pick up a pair of rhinos and bring them alive to London, where they would have had a value of £1,600 (£151,699 in today's money). The first part of the enterprise went well enough. Jamrach Junior successfully acquired the animals and, with forty coolies hauling on ropes, walked them 200 miles to be loaded on the *Persian Empire*, probably at Calcutta. Along with them went food sufficient for 120 days. This should have been enough, but for reasons not explained the voyage was protracted far beyond its normal duration and the animals starved. 'The poor things were reduced to such extremities that they ate sawdust and gnawed great holes in a spare mast,' wrote Buckland.

A French collector, Jean-Yves Domalain, reckoned that ten animals died for every one successfully shown in a zoo, a figure that is impossible to verify but which seems unlikely to have been far wrong. Not all the stories were as harrowing as that of Jamrach's rhinos. Buckland gleefully quotes letters from correspondents describing how monkeys in Brazil and Abyssinia fell for the temptations of alcohol. In the Brazilian case the preferred tipple was cane rum; in Abyssinia it was beer sweetened with dates left out in jars for the animals to help themselves. 'Monkeys certainly will get as drunk as men if they get the chance,' observed Buckland's informant, a Mr J. W. Slade.

The American hunter Frank Buck describes a crafty variation of this trick practised by Malayan natives on an orang-utan. They began by leaving a small tub of water under the animal's tree. The orang duly examined this, and then overturned it. For two more days the exercise was repeated. The tub was refilled and the orang knocked it over. On the fourth day came the breakthrough – instead of spilling the water, the orang drank it. Its capture now was all but assured. After a few more days of water, the Malays began to add increasing amounts of the native spirit *arrack*, until eventually the tub was filled with neat alcohol. After downing this, the orang lurched around like a music-hall drunk, beat his tree with a stick and fell down insensible. 'When he came to, some hours later, he found himself neatly crated at Jesselton [now Kota Kinabalu] awaiting shipment to Singapore.'

But Domalain's Law is no respecter of ingenuity or patience. Frank Buck bought the giant orang – the biggest ever caught – from its captors and sailed off with it for San Francisco and the Ringling Brothers Circus. Knowing his business, Buck took good care of his investment, which was given a roomy cage and fed on carrots, sweet potatoes, bananas, sugar cane, boiled rice, raw eggs and bread. But knowing his business also meant that Buck could face up stoically to disappointment. With about five days of the voyage remaining, the orang went down with dysentery. The ship's doctor struggled to inject a serum but the patient – ill-tempered at the best of times – snapped off all the needles. That night the Rajah of All Orangs, as the Malays had called him, lay down and died.

This was the way it had always gone; the way it always would go.

Like any other traded commodity, animals were bought and sold in a competitive market with dominant market leaders. The earliest and greatest of these was Carl Hagenbeck of Hamburg, born in 1844, the son of a fishmonger who also bought and sold wild animals. It was enterprise on an epic scale. According to Eric Baratay and Elisabeth Hardouin-Fugier in their exhaustive history of world zoos, in the twenty years from 1866 Carl Hagenbeck shifted 700 leopards, 1,000 lions, 400 tigers, 1,000 bears, 800 hyenas, 300 elephants, 70 rhinoceroses, 300 camels, 150 giraffes, 600 antelopes, tens of thousands of crocodiles, boas and pythons, and more than 100,000 birds. Hagenbeck might have been a big player – Phineas T. Barnum was one of many bulk-buyers who depended on him – but he was still only one dealer in a worldwide market that reaped animals like corn. And of course you cannot have corn without chaff. If you multiply the live deliveries by the Domalain factor of ten, then the true cost in lives rises from the epic to the biblical.

Journeys were dangerous, arduous and long. Seasick animals kept in small cages buffeted and drenched by rough seas had only the slimmest chance of survival. 'Tossed about without protection for their claws,' wrote Eric Baratay and Elisabeth Hardouin-Fugier in *Zoo: a History of Zoological Gardens in the West*, 'big cats tore themselves to ribbons and bled to death, or put their own eyes out.' But the casualty rate only added to the value of specimens that were delivered successfully. For a consignment of three African elephants in 1870, an American dealer paid Hagenbeck £1,000 – the equivalent of £100,800 in 2012. Hagenbeck at first thought he had made a pretty good deal. 'But it seems I was wrong. For my American friend took the animals to his own country and sold them for £1,700, £1,600 and £1,500 respectively.' At 2012 values, the dealer's £3,800 profit stacks up to £383,040.

But Hagenbeck reckoned himself to be more than just a money-maker. 'My enthusiasm for my own calling originated more, if I may say so, in a love for all living creatures than in any mere commercial instincts . . . I do not intend to imply that I have not also had an eye to the main chance; but I can, I think, say with perfect truth that I am, and always have been, a naturalist first and a trader afterwards.'

There are good reasons to raise an eyebrow at this generous self-assessment, but Hagenbeck is entitled to some credit for thinking ahead of his time. The zoological park he founded at Stellingen has good claim to be regarded as the first modern zoo. Unlike others of the time it allowed animals to move about with relative freedom in enclosures that bore some resemblance to their natural habitats. 'I desired above all things, to give the animals the maximum of liberty. I wished to exhibit them not as captives, confined within narrow spaces, and looked at between bars, but as free to wander from place to place within as large limits as possible, and with no bars to obstruct the view and serve as a reminder of the captivity.' Artificial mountains were thrown up for chamois and ibex. 'Wide commons' were provided for animals of the plains, and glens for the carnivores, kept apart from the public by trenches. Modern zookeepers may have improved upon this example, but none has ever expressed a more enlightened view.

To modern ways of thinking, Hagenbeck's opinions on animal intelligence could seem anthropomorphic, but at least they encouraged compassion. 'Brutes, after all, are beings akin to ourselves. Their minds are formed on the same plan as our minds; the differences are differences of degree, not of kind. They will repay cruelty with hatred, and kindness with trust.' This led him to reject the old, barbaric methods of training circus animals through threat of pain. He records with revulsion the sight of four 'trained' lions offered for sale at auction in London whose whiskers had been scorched off and who were 'frightfully burned about their mouths'. By example at his own circus in Hamburg, Hagenbeck taught the world's showmen that rewarding an animal was a better way to secure its obedience than punishing it for error. Typically, he did not underestimate the importance of his own achievements: 'There is probably no sphere in which the growth of humanitarian sentiment has been more striking than in the treatment and training of performing animals.'

Typically, too, he did not neglect the bottom line. He looked for profit not just in promoting circus performances of his own but also in supplying trained animals to showmen such as Barnum. Like all enterprises involving the transport of animals, a circus was no place for the risk-averse or the overly sentimental. You might not use the whip or red-hot iron, but the animals were still exposed to hazards against which they had no defence. Hagenbeck himself put together a mixed troupe that included twelve lions, two tigers, several cheetahs and three bears, which he intended to present at the Great Exhibition at Chicago in 1893, and which, after months of preparation, made its debut at Crystal Palace in 1891. All the animals then became ill and died later in Germany of the glanders (an infectious disease usually caused by contaminated food or water), which he blamed on 'the bad meat which was supplied by the unscrupulous contractor in England'. What is bad for animals is seldom good for humans either – a fact that would strike worldwide terror in the early twenty-first century when a global pandemic of avian or swine flu was thought to be only a matter of time. In 1892 Hagenbeck's menagerie was hit hard by cholera, which then spread to the unfortunate people of Hamburg. Even this disaster, however, gave him an opportunity to teach the world a lesson. 'How true it is that cholera is spread through the agency of foul drinking water, was clearly demonstrated by the fact that after the veterinary surgeon had ordered the animals to be given boiled water only, no more of them were attacked by the disease.'

But the vision of Hagenbeck as some kind of latterday St Francis still doesn't quite ring true. One might reasonably ask how all this empathy with creatures whose 'minds are formed on the same plan as our minds' would be reflected in his principal business as an international trafficker of wild animals. The answer suggests an ironclad moral constitution as well as an unsuspected gift for understatement. 'Unlike the hunter, who is attracted only by the love of sport, the animal trader goes to work. He goes, not to destroy his game, but to take it alive; and consequently not the least of the difficulties with which he is beset is the discovery of some practicable way of bringing back his booty to civilisation. As a rule, every foot of the arduous journey is attained only at the expense of some loss to the caravan.' But of course the losses begin long before any caravan gets under way. From Hagenbeck's own account we can see the full and bloody explanation for Baratay and Hardouin-Fugier's statistics of mortality.

First, the animals had to be tracked down and caught. For reasons already explained, capturing dangerous species such as elephant or rhino meant pursuing the young and killing the adults that protected them. Not all were cleanly shot. Hagenbeck describes the method traditionally practised by Nubian swordsmen on horseback, which at least had the merit of spreading the danger more equally between the hunters and the hunted. Indeed, the men deliberately invited bull elephants to charge at them. This was as ingenious as it was dangerous, and relied on deep understanding of the animals' behaviour. Success depended on all but one of the hunters riding dark-coloured ponies. Crucially, the last man was mounted on a grey. As Hagenbeck explains, 'The attention of the elephant, whose sight is not good, is attracted by the colour. Upon the grey pony, the mighty creature usually directs his attack.' All then depended on the rider's skill. His job was to flee, but not so fast that the angry elephant lost hope of catching him. He had to keep tantalisingly just out of reach, the bait in the trap. His companions on the dark ponies meanwhile would close in on the elephant from behind. 'Whoever reaches him first springs from his pony, and delivers a dexterous blow with his sword on the left hind leg of the animal, which cuts the Achilles tendon . . . As the elephant hastily turns to avenge himself upon this new enemy, it becomes the turn of the rider who was formerly being chased to stop, dismount and with a similar blow on the right hand leg to lame the animal on the other side, so that he is totally disabled. If the blows have been delivered with sufficient skill and force, the arteries of the hind legs have been cut, and the elephant bleeds slowly but almost painlessly to death.' In the kingdom of the weasel, that 'almost' reigns without peer.

For giraffes and antelopes the technique was easier. By putting entire herds to flight, all the hunters had to do was wait for the calves to exhaust themselves, when they were easily caught and tied. The hunters would take along a herd of goats to supply the young captives with milk, though this would not prevent more than half of them from dying on the journey. Other species such as hyenas and cats were caught in pits or traps. Hagenbeck took particular pleasure in the method for capturing baboons, which reversed the normal practice of focusing on the young. This time the targets were the highly aggressive tribal elders – dangerous, dominant adults who monopolised the food and who, like villains in a fable, would have their greed turned against them. The first job was to persuade a troop to congregate at a place of the hunters' choosing. This involved blocking all but one of their regular drinking pools with thorn bushes, then baiting the last one with *doura*, a primitive form of sorghum which the baboons particularly enjoyed. Then a trap had to be made. This was like a conical native hut, strongly made from stakes interwoven with branches. Once the baboons had settled down, it was carried to the drinking hole, propped up on one side and baited with *doura*. A cord ran beneath the sand from the prop to the hunters' hiding place. 'Then comes the tragedy. A blazing noonday sun drives the thirsty baboons chattering down to their drinking-hole. Some of the biggest males, who have already secured a monopoly of the *doura*, enter the trap, and commence their feast. The hunter awaits his opportunity: it soon comes; a tug on the cord, the trap closes with a bang, and three great baboons are fairly caught.'

So far so good, but luring them into the trap was the easy part. The 'really critical and dangerous part of the performance' was to get them safely out again and render them harmless. The hunters had to move quickly, or the powerful baboons would smash the trap to pieces. First the men would thrust long forked stakes through the walls and pin the animals' necks to the ground. Then the cage would be lifted off and the prisoners trussed. Hagenbeck, as ever, delights in the detail. 'First their jaws are muzzled with strong cord, made of palm strips; then hands and feet arc tied; and lastly, to make assurance doubly sure, the animal's whole body is wrapped up in cloth, so that the captive has the appearance of a great smoked sausage! The parcel is then suspended from a pole carried by two persons, and conveyed triumphantly to the station.' In this case the 'differ-ences of degree' between the animal and the human mind were held to be somewhat extreme. Hagenbeck believed baboons to be exceptionally stupid, as well as unattractively violent and ungenerous to 'their women'.

The policy of kindness which Hagenbeck lavished upon his circus performers in Hamburg did not translate easily to the wilds of Africa. Here pragmatism and practicality held the whip hand over sentiment. You did what you had to. Necessarily at times there was a Cartesian indifference to pain, a supra ethical convenience that was of particular consequence to young hippo-potami and crocodiles. These were not afforded the compliment of subtlety or subterfuge. They were simply harpooned. Sometimes they were killed, and sometimes merely incapaci-tated. In the former case they were left where they fell. In the

A Hagenbeck camel train carrying specimens across south-west Africa.

latter they were captured in the hope that the wound would eventually heal and a live, if lame, animal could be delivered to the customer. According to Hagenbeck, 'no less than three-quarters of the hippopotami formerly brought to Europe used to be caught in this fashion'.

Other species were caught in pitfall traps. The first snag with these was that lions would often eat the catch before the hunters arrived. The second was the difficulty of raising a half-ton baby hippo out of a hole in the ground. 'When these creatures are agitated they break into profuse perspiration, which causes them to become so slimy and slippery that it is difficult to make the noose hold.' The answer was to pass a rope over the animal's forelegs as well as its neck. 'As soon as the noose is fixed in position the animal is hoisted a few inches off the ground, by the combined efforts of about twenty men pulling on the rope. Half a dozen others jump into the pit, and bind together the forelegs and the hindlegs, as also the jaws; for the animals are obstinate and malicious, and it does not do to run any risks with them.'

Many captured animals died in transit

In fact, there are few accounts of men being seriously harmed by captive animals, but many accounts of animals being harmed by men. Hagenbeck describes a caravan setting out from Atbarah in north-eastern Sudan, aiming to cross the desert to an unnamed port on the Red Sea. In preparation for the journey, stalls and yards are packed with young elephants, giraffes, hippopotami and buffalo. Primitive wooden cages are filled with cats, pigs and baboons. Also being lined up are 150 head of cattle, hundreds of sheep and goats, and more than 100 camels. Moving only at night, the caravan creeps along 'like a great snake, across the wide expanse of glistening sands'. The night-time temperatures do not fall much below the 45 degrees Celsius (113 Fahrenheit) of the day, but at least they are spared the glare of the sun. The large animals are driven along by men on foot – two each for an antelope, three for a giraffe and up to four for an elephant. Smaller animals, such as young lions, leopards, baboons, pigs and birds, travel in rough-hewn cages on the backs of camels. At the very centre of the caravan, pairs of camels are harnessed together with poles lashed across their pack saddles. From each pair of poles hangs a large cage containing a baby hippo, and behind each hippo come another six or eight camels bearing the water it will need during the journey. Every day, *in the desert*, each rhino will wallow in a bath stitched from tanned ox-hide. The rest of the menagerie is fed, according to incli-nation, on the milk or flesh of sheep or goat. With agonising slowness they move between waterholes that may be as much as 60 miles apart and are often defended by armed nomads who charge heavily for access. It is what later generations will call a numbers game. As Hagenbeck wrote in *Beasts and Men*:

> However carefully we organise our expedition, it is inevitable that many of our captives should succumb before we reach our journey's end. The terrible heat kills even those animals whose natural home is in the country. The powerful male baboons are very liable to sunstroke, which kills them in half an hour; and any weak point in their constitution is sure to become aggravated during the journey. Whether this is due to the terror and strain which they underwent at their capture, or to being confined in cramped cages, I cannot say. But the fact remains that not more than half of them arrive safely at their destination, despite our utmost care.

It would take five or six weeks for such a caravan, or the remains of it, to reach the sea. From here, still daily reducing in number, it would be ferried by steamer to Suez, whence it would be either trans-shipped to Europe by vessels en route from India or the Far East, or sent by train to embark from Alexandria to Trieste, Genoa or Marseilles. Hagenbeck himself preferred the railway, even though on one such journey three elephants were killed by rats gnawing their feet. The entire journey from Atbarah took three months. Ironically it was only at the very last stage – elephants or giraffes being marched 8 miles from the docks to London Zoo, for example – that the newspapers took any interest.

Other trips were even worse. Bringing wild foals to Hamburg from Mongolia took eleven months, and cost the lives of twenty-four of the fifty-two animals that embarked. Of more than sixty wild sheep making the same journey, not a single one survived. Despite all this, Hagenbeck's perception of other species as 'beings akin to ourselves' does have a strange consistency. By simple logic, if animals are akin to us, then we must be akin to them, and in the diversity of our own species we should find as much to amaze us as in all the oddities of the jungle. Thus did Hagenbeck hit on 'a brilliant idea'. Alongside the animals he would exhibit in his zoo and travelling shows, he would display exotic *people*. He began with a family of Lapps, whom he shipped to Hamburg with their reindeer in 1874.

> The first glance sufficed to convince me that the experiment would prove a success . . . On deck three little men dressed in skins were walking about among the deer, and down below we found to our great delight a mother with a tiny infant in her arms and a dainty little maiden about four years old, standing shyly by her side. Our guests, it is true, would not have shone in a beauty show, but they were so wholly unsophisticated and so totally unspoiled by civilisation that they seemed like beings from another world. I felt sure that the little strangers would arouse great interest in Germany.

He was right. 'All Hamburg comes to see this genuine "Lapland in miniature"', set up in the grounds behind

Hagenbeck's house in Neuer Pferdemarkt. He attributed its success to the exhibits themselves having 'no conception of the commercial side of the venture', so it did not occur to them 'to alter their own primitive habits of life. The result was that they behaved just as though they were in their own native land, and the interest and value of the exhibition was greatly enhanced.'

The great virtue of Hagenbeck's account is what seems now like almost reckless honesty. If he expected to be judged, it was by people who shared his passion for enterprise. Just as it raised no objection to the exploitation of beasts from Africa, so contemporary thought presented no obstacle to the human parallel. Efficacy was the test, not ethics. 'My experience with the Laplanders taught me that ethnographic exhibitions would prove lucrative; and no sooner had my little friends departed than I followed up their visit by that of other wild men.'

These included first Nubians and then Greenland Eskimos, who were displayed in Paris, Berlin and Dresden as well as Hamburg. Then came 'Somalis, Indians, Kalmucks, Cingalese, Patagonians, Hottentots and so forth'. They were soon worth more to him than elephants. 'Towards the end of the seventies, especially in 1879, the animal trade itself was in an exceedingly bad way, so that the anthropological side of my business became more and more important.' The high point came with his great Cingalese exhibition of 1884, when a travelling caravan of sixty-seven men, women and children with twenty-five elephants and many different breeds of cattle caused 'a great sensation' in Europe. 'I travelled about with this show all over Germany and Austria, and made a very good thing out of it.'

There was innocence as well as calculation in Hagenbeck's thinking, and it may be wrong to convict him of anything worse than naivety, or of being a man of his time. As usual, ironies are not far to seek. Eighty-three years later, the zoologist

Human exhibits – a Greenland Eskimo and his family, displayed by
Hagenbeck at his zoo

Desmond Morris, a former curator of mammals at London Zoo,
would famously write *The Naked Ape*, a uniquely unemotional
review of humankind as an evolving natural species. It sold by
the thousand, and made Morris a household name. As popu-
larisers tend to do, he raised hackles in the scientific community,
but his evolutionary approach to human behaviour caused no
great offence to liberal opinion. Indeed, with its absence of value
judgements it rather chimed with it. By contrast, we look back
upon the scientific anthropology of Hagenbeck's time with something close to revulsion.

The nineteenth century was *the* great age of discovery and
classification, when specimens poured into zoos and museums.
In London, the British Museum began its system of registering

new specimens in 1837, and within a decade it was receiving more than a thousand mammals a year. To a greater or lesser extent, the same thing was going on all over the world. Everywhere, clarity struggled with confusion. Identical species might be given different names by different scientists, or similar names with different spellings, and multiple groups of similar species might be recorded under a single name. It was the age of the enthusiast, when *amateur* was still a term of approbation applied to men of intellectual curiosity. Despite Darwin (himself a Christian), scientific thought was still channelled through faith in God. The superintendent of the Natural History Museum, Richard Owen, rejected Darwin's theories and fought with Darwin's friend Thomas Huxley over what the museum should actually represent. Huxley believed it should be what it has since become, a specialist institution devoted to scientific study within which only a fraction of the collection could be exposed to public view. Owen believed it should be annexed to the Old Testament, setting out with all due wonder and humility the miraculous entirety of God's divine Creation. Even then, the antique quaintness of Owen's view opened him to ridicule in press and parliament, but until his retirement in 1883 the museum would kneel more readily to God than it did to Darwin.

In a way it was of no consequence. Whether the inspiration came from God or from the genius of those created in his likeness, there was a hunger for natural science that gripped the imaginations of educated men and women. Throughout the civilised world and beyond, they came forth in multitudes. In England, bewhiskered physicians, learned doctors and reverend gentlemen toured the countryside measuring, classifying and, in later years, photographing everything they saw. Nothing lived that was not labelled, and the inquiry did not confine itself to beetles, orchids and finches. Like Hagenbeck, the inquiring gentlemen soon found themselves as fascinated by their own species as they were by any other. With callipers, rulers and weights they categorised examples of *Homo sapiens* with a zeal that stopped only just short of the specimen jar. They called their science 'anthropometry', and began to speak of 'breeds'.

In 1900, William Z. Ripley's anthropological field guide, *The Races of Europe*, identified among others the Old Black Breed, the Sussex, the Anglian, the Bronze Age Cumberland, the Neolithic Devon, the Teutonic-Black Breed Cross, the Inishmaan and the Brunet Welsh. Like breeds of dog, sheep or cattle, they all had their defining characteristics. Some were dark; others pale. Some had woolly hair; others fair and fine. Some tended to plumpness; others were thin and wiry. All were shown like prize livestock, staring into the distance with empty eyes. To a modern viewer they look more like criminal mugshots – an observation with which I suspect Ripley, then an assistant professor of sociology at the Massachusetts Institute of Technology, would not have been displeased.

He noted with evident approval that members of the aristocracy tended to be blond and tall, whereas the old British types, with their big ugly noses, wide mouths, heavy cheekbones and 'overhanging pent-house brows', were coarse and rugged. Noses in particular could speak more eloquently than their owners. According to Bishop Whately in his *Notes on Noses*, the typical British type was 'anti-cogitative', as if the size of the nose were in inverse proportion to the size of the brain. Ripley believed that in the proportion, moulding and texture of flesh, bone and hair he could read every nuance of a human breed's pedigree and character.

Few if any pent-house brows were raised at this, for his

views fitted snugly within the Victorian mainstream. The august British Association for the Advancement of Science had its own Anthropometric Committee, whose paper of 1883, 'defining the facial characteristics of the races and principal crosses in the British Isles', had been of great use to William Z. Ripley. The benchmark was *Crania Britannica*, a vast survey of ancient British skulls by two professional craniologists, Joseph Barnard Davis and John Thurnam. This had been published in 1856 and, by her own 'very liberal permission and favour', was dedicated to 'Her Most Gracious Majesty, Victoria'. In the manner of the time, Davis and Thurnam combined meticulous record-keeping with wild assertion. For the people-watchers of the Victorian empire, racial classification was not just a matter of physical differentiation – height, weight, pigmentation, shape and size of skull – but of psychological, intellectual and moral values too. Hence the belief that lunatics and criminals, like foreigners, could be identified by application of a tape measure to the frontal, parietal and occipital regions of the skull. 'It would appear,' reported one celebrated Victorian anthropologist, 'that dark eyes and black or very dark hair are more common among lunatics than among the general population.'

All too clearly now we can see where this was heading. In less than fifty years, the hobby-science of Victorian country vicars brought us to Hitler's master race and the greatest catastrophe of modern history. Among the incidental casualties of war was the use of anthropometrics as a study of living populations. No longer could we talk innocently of 'race' and 'blood'. Racial history was not just politically incorrect; it was politically unthinkable. It was also, for the most part, just plain wrong. *Brutes, after all, are beings akin to ourselves.* For all the suffering he caused, and for all the contradictions of his own example,

Hagenbeck's thinking has a resilient kernel of usefulness. Whether or not we assign moral equivalence to other species, there is virtue in caring. The commodification of animal life, the casual dispensation of unfelt cruelty, kicks open the door to barbarism. If we value the measurable above the infinite, then we lose sight of what it is that makes us human. The wildlife and slave trades of the nineteenth century left scars that have yet to heal. In a sense, Hagenbeck was ahead of his time. Writing of circus animals, he observed, 'it is impossible to achieve by ill-treatment one-hundredth part of what can be done by humane and intelligent methods'. It could as easily have been a stricture on the treatment of children in the London of Henry Mayhew and Charles Dickens, or on the Europe of Hitler and Stalin. In searching for the golden mole, I feel, I am connecting with a thread of what ought to be common humanity.

On that front, there are two new points of interest. First, my philosophical friend Oliver Riviere has sent me a picture of a 'golden mole' that he has trapped in his English garden. It is undoubtedly golden, it is undoubtedly a mole, and it is a puzzle. The European mole, *Talpa europaea*, is typically dark slate in colour, but this one is like fine-cut orange marmalade. How to explain? I assume it's an albino.

More to the point, passing through London I go to look at the stuffed animals in the Natural History Museum. Faded in their cabinets, they are kept now more as historical curiosities than the unique specimens they were when Selous and his contemporaries first took aim at them. A short-beaked echidna shares its case with a duck-billed platypus. There are anteaters and armadillos, a vampire bat, a flying lemur, a pangolin and a hyena. The big cats – cheetah, lion, tiger, jaguar, snow leopard – have faded over time into gaunt, sepia-tinted memories of themselves. Apologetic notices explain that the museum no longer collects skins for taxidermy. But never mind. Here among all the giants and curiosities of the jungle, unregarded and unphotographed by anyone but me, is an unprepossessing scrap of sand-coloured fur. The giant golden mole, *Chrysospalax trevelyani*, looks nothing like itself. It is tail-less, coarsely furred rather than conventionally moleskinned, with no ears, eyes or visible feet. The process of taxidermy has left it with an improbably shiny, joke-shop nose. Overall it looks more like a novelty slipper than anything that might once have had breath in its lungs. The word 'giant' is not misplaced. *C. trevelyani* is roughly twice the size of other species, including the Somali. The museum label explains that golden moles eat worms and other soil-dwelling animals. They are active in the rainy season but may become dormant during the dry or cold season. Looking it up afterwards, I find that the giant golden mole was first described by the museum's newly appointed keeper of zoology, Albert Günther, in the Proceedings of the Zoological Society of London in 1875: 'Mr Herbert Trevelyan has presented to the Trustees of the British Museum the skin of a new species of *Chrysochloris* which he distinguished by its gigantic size . . . He obtained it from a Kaffir who accompanied a shooting-party in the Pirie Forest near King William's town (British Caffraria), and believes that it must be very scarce or local, as none of his companions had ever seen another specimen. Unfortunately the skull has not been preserved; otherwise the skin is in a most perfect condition. I name this species after its discoverer . . .'

He reported that *Calcochloris trevelyani* (since renamed *Chrysospalax trevelyani*) was nine and a half inches long. 'The colour and quality of the fur reminds one of an Otter; it is moderately long, rather stiff, and of a deep chocolate brown colour, with a dense whitish under-fur. Margin of the lips white. On the abdomen the fur is less dense and shorter; and patches of the whitish under-fur are visible in the posterior parts of the abdomen. Muffle flat, projecting as in the other species, but comparatively narrower. Claws whitish; the inner and outer of the fore foot very conspicuous. The third twice as strong as the second. No trace of an opening for the eye or ears, or of the tail can be discovered.' Only the first sentence and the last can be verified from the time-worn specimen whose picture now adorns my mobile phone.

I enjoy the diversion, but we are not quite done with Hagenbeck yet. An incident in the 1880s, after a rogue elephant nearly killed a keeper at his zoo, laid bare all the moral du-plicities of the *soi-disant* naturalist. There was no place for sentiment. 'At any moment a fatal accident might occur; there was no help for it, the monster must be executed,' he writes. *Beings akin to ourselves?* Well, not as akin as all that. It is difficult to imagine a modern zookeeper coming to such an unhappy conclusion; and even harder to imagine one colluding in what happened next. If unsentimentality was one side of the pragmatist's coin, then opportunism was the other. Some people might see an elephant's death as tragedy. Not many would join Hagenbeck in seeing it as just another opportunity to turn blood into money. The chance came during a trip to England, when he mentioned the condemned elephant to the taxidermist Rowland Ward, who then came up with 'a most original proposition'.

'If the elephant were to be had cheap, he said he would will-ingly buy him from me, for he believed he could easily find a "sportsman" to whom it would be worth fifty pounds to be able to say that he had once shot an elephant!' Sure enough, hot-foot to Germany came 'a certain Mr W . . . for the purpose of shooting big game in my Zoological Garden'. The elephant was driven into the yard and tethered to the wall for the hunter to bag his trophy.

All was in readiness, but the hero of the story did not appear. What could have happened? We waited for an hour, and then, as the sportsman still did not arrive, I hastened into the town to remind him of his engagement. I found him and brought him back to the hunting-ground, and at twelve o'clock we gathered around to see the hunter slay his game. The gentleman had brought along his arsenal, but now that he was in sight of the victim the sporting ardour seemed to have unaccountably left him. He fingered his murderous weapons, but did not fire the fatal shot. Presently one of my travellers, who happened to be present, offered to fire the shot, but this the owner of the elephant refused to allow.

The story ends with a cynicism so extreme it becomes bleakly comical. The elephant is ushered back into its stable, a noose is placed around its neck, and six men haul on a rope to hang it from a beam. As an epitaph to an era in which knowledge played hide-and-seek with understanding, it is unimprovable.

But there is something else about Hagenbeck that connects more appealingly to the present. Even now, as more and more species are being hustled to the brink, zoologists still cling to the optimism of their boyhoods. Somewhere, species not seen for decades, even centuries, must be living out their secret lives. Somewhere – in the depths of a Highland loch, in the high passes of the Himalayas, at the sodden heart of a Madagascan jungle – creatures yet unimagined may wait to be discovered.

For all his hard-headedness, Hagenbeck never lost sight of the dream.

Independent reports from reliable witnesses, supported by cave paintings, convinced him that the swamps of Rhodesia contained 'an immense and wholly unknown animal . . . half elephant, half dragon. . . . From what I have heard of the animal, it seems to me that it can only be some kind of dinosaur, seemingly akin to the brontosaurus. As the stories come from so many different sources, and all tend to substantiate each other, I am almost convinced that some such reptile must still be in existence.' An expedition sent to find it was inconvenienced by fever, vast areas of swampland and 'bloodthirsty savages'. It returned empty-handed, but the boy in Hagenbeck would not be quiet.

> Notwithstanding this failure, I have not relinquished the hope of being able to present science with indisputable evidence of the existence of the monster. And perhaps if I succeed in this enterprise naturalists all the world over will be roused to hunt vigorously for other unknown animals; for if this prodigious dinosaur, which is supposed to have been extinct for hundreds of thousands of years, be still in existence, what other wonders may not be brought to light?

Though the 'wonders' of his imagination were coloured by a showman's lust for size and ferocity, and though the Rhodesian swamp-monster would never be found, serious-minded zoologists ever after have been fired by the same unquenchable optimism, and surprisingly often have been rewarded by the reappearance of 'living fossils' or by entirely new species. In my own small way, I am in the grip of it myself. It is not a monster that I seek, but only a tiny scrap of evidence that will prove the existence of the world's most elusive mole. People still ask me why, and my answer echoes the oldest enticement known to man. *Because it is there.*

CHAPTER FOUR

Werewolf Seized in Southend

A hundred years after Hagenbeck's death (he died, aged sixty-eight, on 14 April 1913) we stand atop a mountain of accumulated historical data. From this high vantage point we have clear sight of how animals are feeling our influence. The evidence is consistent and unignorable; the conclusions almost too painful to articulate. 'It makes me ashamed to be of the same branch of biology,' said the late Ian Nairn in a film he made for the BBC in 1973. What had inflamed him – brought him almost to the point of tears – was the demolition of a Victorian church. How much more shaming, then, is the destruction of a work of nature? It seems, or *should* seem, almost inconceivable that, like historic churches, libraries and market halls, species cannot be saved without a fight. This very day, the Worldwide Fund for Nature (WWF) has placed an advertisement inviting people to 'adopt' a snow leopard. This elegant native of central Asia has had a hard time of it. Rapaciously hunted for its fur in the old Russian republics, persecuted by farmers, it has declined by 20 per cent in just two generations. Beyond question, it needs help. But so do thousands of others. The giant panda, Marley's golden mole, the black rhino (recently declared extinct in West Africa), the brownstriped grunt . . . Their names alone would fill many pages of this book. The

IUCN *Red List* contains 5,488 species of mammal. Seventy-six of these are extinct with two more gone from the wild; 505 are vulnerable, 448 endangered and 188 critically endangered. Others are 'near threatened' or 'data deficient'. Only 3,110, for the moment, are of 'least concern'.

One of these, unsurprisingly, is the ubiquitous red fox, *Vulpes vulpes*, which has coexisted with humans since the Ice Age. Like no other, it has clung to life with a tenacity that invites all kinds of anthropomorphic projections – it is intelligent, resourceful, limitlessly adaptable, the possessor of unsurpassable beauty. Despite its modest size, the sight of one always quickens the pulse – especially in England, among whose much-reduced native fauna it reigns as undisputed top predator. It is important also because of what it represents in the eternal processes of negotiation between man and beast. It has the widest range of any land mammal on earth, right across the northern hemisphere from the Arctic Circle through North Africa and Central America to the Asiatic steppes, as familiar in Afghanistan, Mongolia and Bangladesh as it is throughout Europe. You would think that 10,000 years would have been enough for us to get used to each other. Yet no other animal, save perhaps the rat, has rubbed so abrasively against its neighbours. Its role in fiction is as old as story-telling – fifty-one of Aesop's fables involve foxes, and children's literature is full of their misdemeanours.

So, too, are some of Britain's more excitable newspapers. As foxes range ever deeper into towns and cities, so the stories darken – no longer the moral fables of Aesop, but cautionary tales scrawled in blood. There is fear in the streets. At one end of the emotional spectrum the fox is a lovable rogue, to be held in awe, welcomed into our wipe-clean, over-manicured lives as our closest contact with the truly wild. At the other, it is a sly, disease-ridden sadist. (Scarcely believable as I type this, a foxhunt trots jauntily past my office – example par excellence of our jangled attitudes.) I am reminded of the front page of England's *Sun* newspaper from 24 July 1987, which reported that a prisoner at an Essex police station had '**SNARLED** and howled with his lips curled back, **CROUCHED** on all fours, **FOAMED** at the mouth and **LEAPT** at police with his hands and fingers rigid in the shape of claws'. Headline:

WEREWOLF
SEIZED
IN SOUTHEND

This is a classic of the 'man bites dog' genre, news simply because it's so preposterous. In the same way, it is the very rarity of fox attacks on humans that ensures big headlines when they occur. As with murder, there is a perverse logic at work. Incidences are too few to translate into a calculable risk, but it is their very rarity that puts them in the news and creates the illusion of danger. A cornered fox may bite if it feels threatened – typically it will deliver a cautionary nip and back off – but it won't go looking for trouble. If you examine a fox skull, you'll find it's small and delicate, with teeth that seem surprisingly insubstantial until you remember that they evolved to eat mice. Foxes are only slightly heavier than cats, and usually come off worse in fights. For this reason they tend to steer clear of them. The average home territory of an urban fox is half a square kilometre, which means that it overlaps with some 250 pet or feral cats. In Bristol, which until the mange epidemic of the 1990s had the highest density of foxes anywhere in the world, each fox on average is responsible for 0.17 cat-deaths per year – negligible when compared to the numbers ill-met by cars. Most foxes never kill one at all. Violent interactions with humans are exponentially rarer even than this – a handful a year, but with every drop of spilt blood making news. Compare this with dogs. *Every day* in the US, a thousand people need emergency treatment for dog bites. In England it's more than five thousand a year, but it takes a dead child to turn one into a story. As I have discovered, there is no surer way of guaranteeing hate mail in Britain than to write something antagonistic about dogs, their breeders or owners; and no easier way to curry public favour than to write something prejudicial about foxes.

In a way, this is the least of the ironies. Thanks to television, most people now can recognise many more animals than were known to our nineteenth-century forebears, for whom 'nature' was the visible hand of God. No one without top hat, frock coat and beard would have known an echidna from a pangolin, or a tenrec from a hedgehog, or been certain that a whale was not a fish. Now we are both more familiar with and more remote from our co-inhabitants of the planet. Wildlife has been put on the other side of a screen, odourless and with an orchestral backing-track. It's a glimpse, not an experience, an incidental contributor to the urban mindset that puts people and wildlife into a divided world of ours and theirs. The fox here stands for the whole of nature. We love it, but we don't want to share our space with it. It is at once ubiquitous and alien, to be feared as well as loved. Every time a fox, rightly or wrongly, is blamed for some urban outrage, the cry goes up for a cull. The populist mayor of London, Boris Johnson, was at it again in early 2013 when a baby had its finger bitten off.

No amount of science will slough off the ancient myths. Many people believe, and newspapers report as fact, that foxes hunt in packs. But they don't. Why would they? A mouse is not a wildebeest. An animal evolved to hunt small rodents doesn't need the help of others. A more recent myth, eighty years in the making, is that town foxes are somehow different from country ones – degenerates that have forgotten how to hunt, and which survive by raiding dustbins. We chuckle now at the great eighteenth-century clergyman–naturalist Gilbert White for taking seriously the idea that swallows spent the winter at the bottom of ponds. If White could revisit us now, he'd chuckle right back. Given harder currency by the electronic swirl of twenty-first-century media, our own myths are no less risible, and are harder to defend because we have the evidence that proves them wrong.

The fact is that urban and rural foxes are exactly the same. All foxes are opportunistic omnivores that eat as they find. It is true that diets differ, but that's what it means to be omnivorous. You could say the same of humans. A London fox typically will eat 24.1 per cent meat, bones or fat (including food left out for it, and scavenged pet food). This it will beef up with 14.4 per cent wild birds and 13.1 per cent small mammals. Earthworms (12.2 per cent) are more important than pet birds (5.8 per cent) and pet mammals (2.9 per cent) added together. Bristolian foxes eat more meat (32.6 per cent) than the Londoners, but a much lower proportion (0.9 per cent) of pet mammals. What *none* of them does is raid dustbins. Years ago, the Mammal Research Unit at Bristol University monitored more than 5,000 household bins and found that only 2.7 per cent of them were regularly disturbed by foxes. Even that turned out to be an exaggeration, as surveillance revealed that many of the visitors were not foxes at all but dogs and cats. Since then the old-fashioned dustbins have been replaced by wheelie bins inaccessible to anything smaller than a polar bear.

After fear of attack comes dread of fox-borne disease – specifically sarcoptic mange, which is triggered by the same mite that causes scabies in humans. At least one British newspaper has linked it to the 'rotten diet' supposedly scavenged from garbage tips, and it is a spectre commonly raised by pest-control companies touting for business. Mange is a hazard for all foxes, town and country alike. In the late 1880s an epidemic so depleted the English counties that foxes had to be imported from Europe for the hunts (a fact not much mentioned by those who argue that hunting is all about pest control). The first modern case in Bristol was in 1994 and involved an animal that had spent the winter outside the city and brought it back in the spring. The resulting outbreak reduced the population by more than 95 per cent. But mange has nothing to do with garbage. The mite is passed on by contact with other foxes or dogs. The more rational fear is of the roundworm *Toxocara canis,* which can cause blindness and is spread by faeces. But even in areas heavily used by foxes, faeces are extremely difficult to find. Students involved in the Bristol study often could find none at all, even in heavily populated areas. Not so with the dog, whose promiscuous faecal gifts have been responsible for every known case of toxocariasis in humans. For the reasons already given, I hesitate to mention this in public.

It is not just this that makes culling senseless. Overall, leaving aside the mange epidemic, Britain's fox population has remained remarkably stable. Every year before breeding starts, the number of adults stands at around 250,000. A year later it will be the same – 425,000 cubs will have been born, and 425,000 foxes died. The reason is that there is a finite number of territories, and this is why shooting is a waste of bullets. If you killed every one you saw, within three or four days it would be replaced by another – then another, and another for as long as you could afford the ammunition or had the energy to go on pulling the trigger. There are always itinerants looking for new or better territories. It's the same in the country, where the economic case for fox control was seriously weakened in the last century by the industrialisation of egg and chicken production, which put most of the nation's hens beyond reach. On agricultural land foxes could even be counted a blessing. It has been estimated that a single fox in its lifetime can save farmers between £150 and £900 in damage by rabbits.

These are just some of the myths, misunderstandings and prejudices that attend a single common species living close to man – just one small peak in a mountain range of stupidity. We can smile at, or be disgusted by, the ignorance or unfeelingness of men who cut belt-loops in a living hippopotamus, hang an elephant or feed a panda on grass. But comparisons with the nineteenth and early twentieth centuries may now look a little less weighted in our favour. The twenty-first century is a glass-house from which stones should be cast only with the greatest caution.

One inviting target for retrospective judgement is Frank Buck, the extravagant Texan who became a dominant figure in the wild animal trade after the Hagenbeck firm was crippled by the First World War. For all his self-promotion, however, Buck is a hard man to dislike. The world loves an adventurer, and it certainly loved Buck. In his youth he was an apprentice safe-cracker and bar-room brawler, but he grew up to become one of the great all-American heroes. He may be forgotten now, but in the 1930s his celebrity equalled the likes of Charles Lindbergh, Babe Ruth and Gene Tunney. His books, written in a hard-boiled style closer to Raymond Chandler than to Rudyard Kipling, were bestsellers, his filmed exploits were cinematic blockbusters. A powerfully built man with an Errol Flynn moustache, Buck looked as well as acted the part – exactly the kind of man you would expect to come out on top after a bare-handed struggle with a cobra. His bravery usually was calculated to stay just on

The Texan swashbuckler Frank Buck, who stocked America's zoos and
starred in the circus

the right side of recklessness, and he usefully introduced
Americans to many exotic creatures they had never heard of or
suspected.

'Nothing delights me more,' he wrote, 'than to place under
the nose of an eminent zoologist a bird or a beast or a snake he
has never seen before . . .' Trapping or buying to order, he kept
American zoos and circuses stocked with man-eating tigers and
other crowd-pleasing exhibits and performers. The most crowd-
pleasing exhibit was usually himself – astride an elephant, he
was a featured act in the Ringling Brothers' travelling circus,
and his books were fast-moving chronicles of derring-do.

No zoo now can offer quite such tremors of excitement as
the characters in Frank Buck's many personal dramas. Provenance

is unemotionally described in terms of captive-breeding programmes, which in every other sense – conservation of species, welfare of individuals – is a vast improvement over the bad old days of plundering the wild. But there is no hint of danger; nothing to pump the adrenalin. Captive-bred animals might resemble their wild counterparts in every physical particular, but they have been stripped of their mythology and of their exotic and dangerous glamour. Public interest is no longer pricked by adventurers' yarns or by the imagined hot breath of yellow-eyed man-eaters.

All the stimulus, and most of the gratification, is now delivered by the camera. Through modern miracles of information technology, we witness the pitiless battles for food, mates, territory, survival – fights between elephant seals so monstrously violent that they seem to shake the house; the daily carnage of the African plains; orcas reddening the sea. But it leaves a gap, an unfulfilled appetite for the thrill of man against beast. It is a peculiar corner of the human psyche that two thousand years ago was gratified by the Roman amphitheatre, and which is now served by exhibitionists on television leaping on to the backs of crocodiles. The most famous of these was the Australian Steve Irwin, whose 'Crikey, he nearly got me!', uttered at critical moments during his championship bouts with crocs, Komodo dragons and snakes, became a catchphrase among his multitude of fans. But Irwin himself had no illusions about what drew people to watch him. As he told ABC television: 'Steve Irwin's all pretty interesting on the telly or in the movie and that, but by crikey, it's great when he gets bitten. Now and again I do get bitten. But I haven't been killed. And it's that, you know, that sense of morbidity that people do have. There's no use sticking your head in the sand and going, "Oh, no, they're only here because, you know, I

talk well." Nah, man, they wanna see me come unglued.'
Off-camera, at the Great Barrier Reef in September 2006, a
stingray pierced his heart.

Another who paid the ultimate penalty was the ex-heroin
addict, film-maker and eco-activist Timothy Treadwell, subject
of Werner Herzog's film *Grizzly Man*, who for thirteen summers
lived up close and dangerous with bears in Alaska. On 6 October
2003, his mutilated head, part of his spine, right forearm and
hand were found near the wreckage of his camp, where the
partial remains of his girlfriend, Amie Huguenard, lay half
buried under twigs and dirt. Some of the missing body parts
were found inside a large male bear shot by park rangers; others
may have been in a younger bear, also shot, which was itself
eaten by other animals before it could be dissected. Like Irwin,
Treadwell specialised in the kind of risk that even Frank Buck
would have drawn back from.

For most people leading orderly lives, contact with wild
animals is rare and fleeting. Bird tables and nesting boxes are
compensation for some. But we are always being encouraged
to go further. Excited young television presenters lure animals
ever closer into contact, diluting their wildness in a bath of
nursery sentiment. *Aaah!* This is danger of an altogether
different kind. Throughout history, from the meagre knowledge of animal life demonstrated by the scribes of the Old
Testament (when, as Christopher Hitchens pointed out, God
seemed to have had no idea that he had once invented dinosaurs), misunderstanding is a common thread in human
interactions with other species. Disrupting an animal's behaviour is always risky, and sentimentality is no substitute for
respect. Many people (10 per cent of households in Bristol,
for example) are so pleased to see foxes in their gardens that
they put out food for them. The sensible ones scatter it at a

distance from the house, then watch discreetly from a window. That is as far as kindness should go. Luring foxes to the house, trying to feed them by hand, is literally a confidence trick. To hand-feed a wild animal is to ensure confusion when it confronts a less welcoming neighbour. Many fox 'attacks' begin this way.

Vulpes vulpes nevertheless is one of the world's winners, always flexible enough to adapt to change. The sufferings of the snow leopard and black rhinoceros are more typical. Environmental deprivation can be surprisingly easy on the eye. Two years ago I stood on a high bluff overlooking a raw slab of sub-equatorial Africa in the rainy season. The vegetation was of the kind that, in the language of television, could only be called 'lush'. Birds in a multiplicity of sizes and colours called and responded like a sonata for woodwind – oboe, flute, piccolo – while insects gnawed at my burning neck. Africa never lets you forget how vulnerable you are.

Two-hundred-year-old sterculia trees testified to a primeval landscape in vibrant good health, Gaia absorbed in self-love, while rainclouds layered above the hills promised more of the same. The conical thatched roofs of village houses were a frail, distant archipelago in a trackless ocean of green – men, women and children subsumed by nature like termites. It looked utterly, unimproveably right.

And yet it was wrong – completely, utterly wrong, the eye deceived, as it so often is, by beauty. I should have been looking at dense riverine forest, not at this inviting grassy plain. The huts should have been hidden away in clearings known only to those who were born in them, and the sterculias lost among trees as tall as themselves. They were there only because it was taboo to harm them – the local people, who know them as *Mudjerentjes*, believed the rains would stop if they cut them down. The glorious panorama that filled our lenses simply should not have existed.

This part of central Mozambique has had it harder than most. Sixteen years of civil war cost many thousands of lives. The wildlife – once a bankable asset in a thriving tourism industry – was killed to feed soldiers, and the forest was cut down for fuel to cook it on. What the fighters had begun, illegal logging gangs and charcoal burners had done their best to finish. You couldn't blame people for taking what they could. Malnutrition is no great friend of wildlife conservation. Add rampant malaria, Aids and patchy medical care often rejected in favour of the chants and potions of witch doctors, and it was a miracle that average life expectancy reached as high as forty. For protein at this time of year the people had termites, ants and caterpillars. And yet there is another misperception into which your eye might lead you. The alienation of the landscape from its oldest inhabitants is not an event: it is a process.

It is often said that the first significant work of conservationist literature, the first plea for wildlife against malign human intervention, was Rachel Carson's *Silent Spring*, published in 1962. This great and brave book did much to launch the environmental movement that grew in the decades that followed, but the echo it picked up – of nature battered and bruised – was very much older. In 1864, Abraham Lincoln's United States was fighting the Confederates in the American Civil War. The future Chancellor of Germany, Otto von Bismarck, was still Minister President of Prussia; Napoleon III was emperor of France, Tsar Nicholas I and Franz Joseph I bestrode their empires in Russia and Austria. In London, Her Britannic Majesty's prime minister was Lord Palmerston. In that same year Charles Dickens published *Our Mutual*

Friend, and Jules Verne *A Journey to the Centre of the Earth*. Nathaniel Hawthorne and John Clare died. Henri de Toulouse-Lautrec, Richard Strauss and Alois Alzheimer were born. The Royal Navy launched its biggest, fastest (and last) wooden warship, *HMS Victoria*, and Isambard Kingdom Brunel's Clifton Suspension Bridge opened in Bristol. A Dutch brewer called Heineken opened for business, and James Robertson started making marmalade in a room behind his grocer's shop in Paisley. Also in that year, nearly a century before *Silent Spring*, the American diplomat and philologist George Perkins Marsh published his masterwork, *Man and Nature; or Physical Geography as Modified by Human Action*. He began it with a quotation from Horace Bushnell's 1858 *Sermon on the Power of an Endless Life*:

Not all the winds, and storms, and earthquakes, and seas, and seasons of the world, have done so much to revolutionise the earth as MAN, the power of an endless life, has done since the day he came forth upon it, and received dominion over it.

Like Alfred Russel Wallace, Marsh had one of those cavernous, polymathic nineteenth-century minds with a seemingly infinite capacity for storing facts. In *Man and Nature* he cites references from 210 different publications, many of them written in German, French, Italian, Dutch and even Norwegian. It is not an example that the current writer is remotely capable of emulating. To say that Marsh showed prescience is to understate his achievement by several magnitudes. 'Sight is a faculty,' he wrote. 'Seeing, an art. The eye is a physical, but not a self-acting apparatus, and in general it sees only what it seeks.' What his own eye sought was an unglossed, finely detailed picture of the earth as reshaped by human activity. A photograph shows an apparently large, bespectacled, somewhat pear-shaped man stiffly dressed in frock coat and weskit, holding a silk top hat. He wears a voluminous Abe Lincoln beard and the severe expression of a man not given to frivolity (though that might be accounted for by the need to hold a pose for the photographer). No elder of the church ever looked more forbidding, and he was not shy of handing down judgements. On bad habits, for example: 'I wish I could believe, with some, that America is not alone responsible for the introduction of the filthy weed, tobacco, the use of which is the most vulgar and pernicious habit engrafted by the semi-barbarism of modern civilisation upon the less multifarious sensualism of ancient life . . .'

Ancient life is a constant point of reference, an anchorage for his thoughts. He looks to the Roman empire and sees 'a productiveness of soil of which we at present discover but slender traces'. That bygone fertility, he argues, explains how vast and hungry armies, like those of the ancient Persians, the Crusaders and the Tartars could provision themselves throughout long marches in lands that now 'would scarcely afford forage for a single regiment'. All around him he sees the scars of 'man's ignorant disregard of the laws of nature'. The indiscriminate slaughter of insect-eating birds, for example, has dire consequences for crops and wild plants. Man, he says, 'is everywhere a disturbing agent. Wherever he plants his foot, the harmonies of nature are turned to discords.' The effect is of 'treacherous warfare on his natural allies'. Marsh is one of the first – maybe *the* first – to note the risk of climate-change from deforestation:

When the forest is gone, the great reservoir of moisture stored up in its vegetable mould is evaporated, and returns

only in deluges of rain to wash away the parched dust
into which that mould has been created. The well-wooded
and humid hills are turned to ridges of dry rock, which
encumbers the low grounds and chokes the watercourses
with its debris, and – except in countries favoured with
an equable distribution of rain through the seasons, and
a moderate and regular inclination of surface – the whole
earth, unless rescued by human art from the physical
degradation to which it tends, becomes an assemblage of
bald mountains, of barren, turfless hills, and of swampy
malarious plains.

Human improvidence, he concludes, threatens to reduce
the earth 'to such a condition of impoverished productiveness,
of shattered surface, of climatic excess, as to threaten the
depravation, barbarism, and perhaps even extinction of the
species'. It's the kind of talk which these days would arouse a
chorus of furies on libertarian blogs and websites. Debate 150
years ago could certainly be polarised, but it was religion and
rationalism that were squaring up, not climate science and
libertarianism. *That* conflict had yet to be imagined. Marsh
was forever conscious of his Creator, and of the responsibilities implicit in man's dominion over what the Lord had
provided. God's work, he contended, was everywhere superior
to the works of man:

Not only is the wild plant much hardier than the domesticated vegetable, but the same law prevails in animated
brute and even human life. The beasts of the chase are
more capable of endurance and privation and more tena-
cious of life, than the domesticated animals which most
nearly resemble them. The savage fights on, after he has

received half a dozen mortal wounds, the least of which would have instantly paralysed the strength of his civilised enemy, and, like the wild boar, he has been known to press forward along the shaft of the spear which was transpiercing his vitals, and to deal a deathblow on the soldier who wielded it.

That may be so, but noble savagery is a flimsy defence against malevolent genius. The 'animated brute' had no answer to a 'civilised enemy' armed with guns. The 'terrible destructiveness of man', Marsh said, was exemplified by the hunting down of large animals and birds for those portions of their bodies – often a very small part of the whole – which had commercial value. The wild cattle of South America had been 'slaughtered by the millions for their hides and horns', the North American buffalo for its skin or tongue, the elephant, walrus and narwhal for their tusks; whales for their oil and whalebone; the ostrich and other large birds for their feathers. Already one big marine mammal, Steller's sea cow, *Hydrodamalis gigas*, had been driven to extinction – 'extirpated', in Marsh's language – for the sake of its oil, fat and fur.

Marsh saw that where the sea cow had led, others were certain to follow. Seals, walruses and sea otters were also suffering, and the more valuable fish had been 'immensely reduced in numbers'. Though in fact many mammals had already been forced out of existence by 1864, this was still only the thin end of the wedge. Not even Marsh could have imagined the mass extinction that would be under way by the early years of the twenty-first century, though he had a pretty shrewd idea of the way things were going.

Five hundred years ago, whales abounded in every sea. They long since became so rare in the Mediterranean as not to afford encouragement for the fishery as a regular occupation; and the great demand for oil and whalebone for mechanical and manufacturing purposes, in the present century, has stimulated the pursuit of the 'hugest of living creatures' to such activity, that he has now almost wholly disappeared from many favourite fishing grounds, and in others is greatly diminished in numbers.

The one piece of good news was the silk topper in Marsh's photograph. Hitherto, the favoured material for gentlemen's hats had been the fur of the beaver – a material of high economic value for which the species had been paying with its life. The irony is that what saved the beaver was not concern for its survival – no celebritics paraded their concern, as they would do in later years for leopard, fox and mink – but rather the ephemeral whim of fashion.

When a Parisian manufacturer invented the silk hat, which soon came into almost universal use, the demand for beavers' fur fell off, and this animal – whose habits, as we have seen, are an important agency in the formation of bogs and other modifications of forest nature – immediately began to increase, reappeared in haunts which he had long abandoned, and can no longer be regarded as rare enough to be in immediate danger of extirpation. Thus the convenience or the caprice of Parisian fashion has unconsciously exercised an influence which may sensibly affect the physical geography of a distant continent.

So it goes, as Kurt Vonnegut's Billy Pilgrim might have said. Today another leaflet drops out of my morning paper, courtesy of WWF. This time it asks me to adopt an Amur leopard – an animal, I confess, of which I know little. There is a good reason for this. According to WWF only thirty-five are left in the wild. Even the IUCN returns no results, though when I broaden the search with the single keyword 'leopard', the screen suddenly overflows. No fewer than fifty-four listed species include 'leopard' in their common name. A few are cats. Some are reptiles (leopard fringe-fingered lizard, leopard snake), amphibians (western leopard toad, Las Vegas leopard frog), marine mammals (leopard seal) and fish (leopard sharpnose puffer, leopard-spotted swellshark), so called presumably because they have spots. The swellshark, I learn, is known only from a single specimen caught off Taiwan. Many of the others are endangered. The sad fact is that all the examples in this book could be replaced by others of equal or greater importance. What I have written about mammals might just as well have been written about fish, amphibians, reptiles or plants. Human advance is indiscriminate, an assault on many fronts, and only in their last redoubts do threatened species – the lucky few – hear the bugle-call of the relieving cavalry.

The Amur leopard, I discover eventually, is *Panthera pardus orientalis*, also known as the far-eastern leopard, a vanishingly rare subspecies, long-tailed and thickly coated, that clings to life along the Russian–Chinese border in extreme north-eastern Asia. There is a possibility that others may survive in North Korea, but this is not an area open to scientific inquiry, or sensitive to the fears of conservationists. The Amur leopard has been blitzed by a multiplicity of threats, not least the misfortune of occupying such a fraught political hotspot. Much of its habitat has been lost, and the rest fragmented by fires and logging. Its very rarity makes it a target for poachers, and its carnivorous habit provokes retaliation by deer-farmers. To complete the vicious cycle, its confinement to one small local population means it is being enfeebled by inbreeding.

'The felling of the woods,' wrote George Perkins Marsh, 'has been attended with momentous consequences.' It took a perceptive eye to see that in 1864. Anyone can see it now, but – as always with those called 'doomsayers' – being proved right is a poor sort of consolation. It is not quite too late to listen to those who now echo him. As we begin to feel the unpredicted feedbacks from a disrupted climate, his voice from 150 years ago comes over loud and clear. Our 'limited faculties', he says, are blinding us to the ultimate consequences of our deeds.

But our inability to assign definite values to these causes . . . is not a reason for ignoring the existence of such causes in any general view of the relations between man and nature, and we are never justified in assuming a force to be insignificant because its measure is unknown, or even because no physical effect can now be traced to it as its origin. The collection of phenomena must precede the analysis of them, and every new fact, illustrative of the action and reaction between humanity and the material world around it, is another step towards the determination of the great question, whether man is of nature or above her.

The language may be archaic, but the wisdom is timeless.

My search for the Somali golden mole takes one step forward and half a step back. Learning my way around the laconic, abbreviated intricacies of zoological literature, I discover that the world expert on golden moles is Gary Bronner at the

University of Cape Town, who declares his principal areas of interest to be 'phylogenetic systematics, functional morphology and conservation biology of Africa's endemic golden moles; and the structure and ecology of terrestrial small mammal communities'. He it was who wrote the entry for *Calcochloris tytonis* in my new desktop bible, *Mammal Species of the World*. From Bronner I learn that the discoverer of the Somali golden mole was Professor Alberto M. Simonetta, of the Institute of Zoology, University of Florence, and that his description of it was published in the journal *Monitore zoologico italiano*. Nervously, only half understanding the reference and ashamed of my ignorance, I call the library at the Natural History Museum. Can they help? They can indeed. A day later, all twenty-nine pages of Simonetta's paper, 'A New Golden Mole from Somalia with an Appendix on the Taxonomy of the family Chrysochloridae', dated 30 January 1968, are spread out on my desk.

The story is even better than I had hoped. During the summer of 1964, in a Somali town called Giohar, Simonetta discovered a disused oven in which a family of barn owls – two adults and two fully grown young – had been nesting. 'The floor of the oven,' he writes, 'was covered by a layer of dust, feathers, loose bones and owl pellets about three inches thick.' Some people to whom I've recounted this story have not properly understood what an owl pellet is – not faecal matter, but a regurgitated plug containing the undigested parts of what the bird has swallowed (typically fur, feather, bone, beak and claw). If you want to know what an owl has eaten, then the pellets will tell you. The contents of the Giohar oven, however, looked somewhat less than promising. Trampling by the owls, and the destructive attentions of beetles, meant that most of the material was smashed to pieces and only a small number of pellets were still intact. Nevertheless, with grand scientific impartiality, the debris was all swept up and taken back for analysis in Florence.

Then came the essential stroke of luck. 'While sorting the material,' Simonetta writes, 'I found a right ramus of the lower jaw of a Golden Mole, still articulated with the almost complete temporal part of the basioccipital, of the hyoid and the first two cervical vertebrae.'

In layman's language, this translates as the right-hand side of the animal's lower jaw, with temple and bones of the middle ear still attached. It sounds a lot, but it's tiny. The jawbone was not much more than a centimetre long, a scrap that most people would not even have noticed. Simonetta, however, wanted to know what species it was. Theoretically the identification of a mole from such small remains is easier than it sounds – the crucial distinctions are in the size and proportion of the jaws and teeth, exactly what Simonetta had found. It was a golden moment. Comparison showed similarities but no precise match with any other mole. Simonetta had a new species on his hands.

And new it has forever remained. No other sign of *Calcochloris tytonis* has ever been recorded, alive or dead. As a sample of modern mammalian life, Simonetta's specimen MF4181 was, as I had supposed, the ultimate rarity. But I read and digested all this with mixed feelings. Whereas I was pleased that the specimen had been located, and a trip to Florence would be a pleasure, I was disappointed that it had not given me the satisfaction of a harder chase. The disappointment did not last long. As far as I knew, only one other man in the world – Gary Bronner – had any interest in tracking down *Calcochloris tytonis*, and I emailed him for advice. He was the leading specialist in the field, a scientist of international repute, and years ahead of me in the quest. His reply astonished me. Like me, he had expected to find the mole at the University of Florence, or at the Florence Natural History Museum or Institute of Zoology. As I intended to, he had asked after Simonetta's specimen at each of these venerable institutions, and at each one he had drawn a blank.

The hunt was still on.

CHAPTER FIVE

Penitent Butchers

Brumas was born on 27 November 1949, thirteen days after my own fourth birthday. Son of Mischa and Ivy, and named after his keepers Bruce and Sam, he was the first polar bear to be born and raised in Britain. He was an immediate and lasting sensation. In 1950 he boosted attendances at London Zoo to an all-time record of three million, and inspired a profitable trade in Brumas-themed books, postcards and souvenirs. There was something very odd about it, though. For reasons never properly explained, my use here of the words 'son', 'he' and 'his' perpetuate a bizarre error made in the newspapers and allowed to pass uncorrected by the zoo. Brumas was a *girl*, but for as long as she lived (she died in May 1958) the public went on believing her to be a male.

I suspect that my memory of her/him is more imaginative reconstruction than genuine recollection (I conjure a vague picture of mother and infant on a rock, and a faint smell of fish) but it is indelible. It is one of many cherished memories of the 1950s. This was not just the decade of the 'family values' (board games, side partings and Bisto) that still drive the politics of nostalgia. More importantly, it was the decade in which television sets first began to appear in ordinary homes such as my own. Two genres dominated my early viewing: the heroic adventures of cowboy avengers such as Hopalong Cassidy and the Lone Ranger, and wildlife shows. From the middle years of the decade, the much-parodied husband-and-wife team of Armand and Michaela Denis were a regular favourite with their *On Safari* series. Armand was a burly, bristly moustached, bespectacled and thickly accented Belgian film-maker; Michaela a glamorous blonde English ex-fashion designer. Clamped to their binoculars, they trekked by Land Rover in a never-ending African safari, filming as they went. An early, flickering fragment shows Michaela cuddling a jackal. I discovered only recently that Armand's earlier credits as a director included Frank Buck's 1934 film, *Wild Cargo*. ('Although it may seem as though several incidents in the screen work were prearranged,' said *The New York Times*, 'they are nevertheless quite thrilling, especially when the hunter depicts the ingenious methods by which he traps wild beasts and reptiles.')

On Safari now seems every much a bygone as Hopalong and the Lone Ranger, a faintly embarrassing relic of a colonial age in which African people could be captioned on-screen as 'the Natives'. But there were two other popular shows which, on one young mind at least, would have a deeper and much more lasting impact. One of these was presented by a gawkily hand-some, toothy young man whose cheerful disposition reminded me of my favourite schoolmasters – engaging, brimming with enthusiasm but not didactic. The other was (misleadingly, as it would turn out) a more headmasterly figure – bald, somewhat older than the first, and stuffed with knowledge. The handsome young man was twenty-eight-year-old David Attenborough, just hitting his stride with *Zoo Quest*. The other was Peter Scott, with *Look*.

As Attenborough recalled in a filmed interview in 2007, animal programmes hitherto had been of two contrasting types. In the first type – a kind of prototype *Blue Peter* – live animals would be brought to the studio from the zoo. 'You stuffed them in a sack,' said Attenborough, 'and brought them in the middle of the night in a taxi up to Alexandra Palace, and then hoiked the poor things out on to a table covered with a doormat.' The programmes went out live, and the thrill for the audience was the sheer unpredictability of a bewildered animal with fully functioning bladder and bowels, and a yen for freedom. The risk to the presenter's dignity made it 'great television', but it revealed almost nothing about the animal's nature.

The other strand was the Armand-and-Michaela-type film of creatures in the wild, shot usually in Africa. This was more informative, but it lacked the immediacy of live animals in the studio. Attenborough's brainwave was to combine both strands into a single format. The idea was to travel to remote parts of the world to hunt, film and catch rare species never before seen by the public, and – with a modest echo of Frank Buck – to bring them back alive to London Zoo. Hence the title, *Zoo Quest*. The animals then could be brought to the table to entertain the live audience. The plan was for Attenborough to produce the series, and for his friend Jack Lester, the zoo's curator of reptiles, to appear on screen.

For some reason they elected to start on the west coast of Africa, in Sierra Leone, 1,700 miles north and west of the point where Hanno the Navigator had turned for home 2,500 years earlier. In the days before intercontinental jets, it still took them three days to fly there from London (the first leg in a Dakota) via Casablanca and Dakar. 'Sierra Leone' translates invitingly as 'Mountains of the Lion', but the quarry necessarily had to be a bit less daunting than the fabled king of the jungle. London Zoo in any case had had a lion house since 1876, so *Panthera leo* wouldn't fit the criteria of rare and unseen. But neither, it seemed, was there much else in Sierra Leone that could satisfy Attenborough's desire for 'the ultimate rarity'. The best candidate, it turned out, was a bird. To modern ears, the white-necked rockfowl, *Picathartes gymnocephalus*, doesn't sound too much like the stuff of compulsive viewing, or the springboard for one of the greatest careers in broadcasting history. But so it proved, though it was a triumph that grew out of tragedy. In Africa Jack Lester caught a tropical disease from which he was never to recover, and which eventually killed him. He was able to present only the first episode, after which the producer had to step out from behind the camera and fill the gap. As I write, he's still there.

Peter Scott's *Look* series was of the more traditional studio-based kind, though it also made use of film. The very first programme featured a live fox and launched a series that would make Scott, like Attenborough, a household name. Scott might have been remembered for many different things. He was a distinguished wartime naval commander, an Olympic sailor, British gliding champion and a popular artist whose prints – typically of flighting wildfowl – were the only art that many families hung on their walls. He was also an expert ornithologist. For a full account of his extraordinary life, Elspeth Huxley's biography (foreword by David Attenborough) does a rather better job than his autobiography, *The Eye of the Wind*. Better than Scott himself, Huxley describes how his feeling for nature developed through his passion for wildfowling, and how by the early 1950s his shocking proficiency with the gun had turned him away from killing. His decisive conversion to the preservation of life is one of the reasons why, today, I receive yet another invitation from the WWF to support a species under threat – this time the jaguar, *Panthera onca*, largest cat of the Americas. If I prefer, I could choose instead a giant panda, a polar bear, orang-utan, bottlenose dolphin, Bengal tiger, Asian elephant, black rhinoceros, hawksbill turtle or Adélie penguin, and I could receive a 'gorgeous soft toy' of my favourite species. The populism and ubiquity of wildlife conservation now would astonish the far-sighted few who got it moving.

Throughout recent history, the name Huxley has been one of the most prominent in contemporary thought. Scott's biographer Elspeth Huxley was married to a cousin of the writer Aldous and of the evolutionary biologist Julian, grandsons of 'Darwin's bulldog' Thomas Henry Huxley (a younger contemporary of George Perkins Marsh). It was Thomas Henry, inventor of the word 'agnostic', who objected to Richard Owen's Creationist vision of the Natural History Museum, and who opposed Owen and 'Soapy Sam' Wilberforce, Bishop of Oxford, in the famous debate at Oxford in 1860. Soapy Sam made the mistake of trying to demolish Darwin by scorn. Was it through his mother's or his father's side, he wondered, that Huxley had descended from a monkey? In modern political debate, Huxley's riposte would be called a zinger. He would 'not be ashamed to have a monkey for his ancestor', he told the unfortunate bishop, 'but he would be ashamed to be connected with a man who used his great gifts to obscure the truth'. No smirk has ever disappeared more swiftly from episcopal lips, and literal-minded biblical fundamentalism has seldom taken a longer step backwards.

Thomas's grandson Julian would grow up to share the old man's taste for controversy. He was, for example, an outspoken advocate of eugenics, a far from easy thing to be in the years after the Second World War. He was president of the British Eugenics Society from 1959 to 1962, and dispensed from a great intellectual height opinions on the reproductive excesses of society's 'lowest strata'. Although he was no racist, the liberal consensus now would find these opinions difficult to swallow or even to forgive. As a scientist he could see no reason why selective breeding should not be as improving to *Homo sapiens* as it had been to pigs, sheep and cattle. Population statistics continue to be controversial. Generations of scientists and conservationists have seen, and still do see, overcrowding as the most serious threat to life on the planet. (This was the theme of David Attenborough's 2011 President's Lecture to the Royal Society of Arts, when he quoted Malthus's doomy *Essay on the Principle of Population*, which first sounded the tocsin in 1798.)

But it was not any of this that fixed Julian Huxley in the public mind. From January 1941 he was a panellist on BBC radio's hugely popular *Brains Trust*, and in 1946 he became the first director-general of UNESCO. Crucially, though, above all else he was a biologist. From 1935 until 1942 he was secretary of the Zoological Society of London, where his progressive ideas put him in almost perpetual conflict with the Fellows. Echoing his grandfather's vision for the Natural History Museum, he wanted the zoo to give more emphasis to research and education. He added more informative labelling to the cages, and – in the face of opposition from the Fellows – established the world's first Children's Zoo (officially opened by Robert and Edward Kennedy, the two younger sons of the then US Ambassador to Britain). Always he was driven by a mission to inform and enlighten. He wrote popular books about the zoo and its animals, and launched a monthly *Zoo Magazine*. The Fellows, however, rejected his scheme for a natural history cinema, and blocked his attempt to save money by cutting senior staff. His plan to buy Eric Gill's nude female statue *Mankind* for the society's rural outpost at Whipsnade, on the Dunstable Downs in Bedfordshire, also came to grief. Chimpanzees were one thing, but a kneeling, headless human female *au naturel* was, in the Fellows' view, not something to be set before visitors to a zoo. (The statue subsequently was acquired by the Tate Gallery, and at the time of writing is on loan to the Victoria and Albert Museum.) The Council was also outraged by Huxley's freelance activities as a broadcaster, writer and lecturer. After three months' unpaid leave lecturing in the United States, he resigned in 1942.

By 1960, at least seventy modern mammals, described and known to science, including species of gazelle, deer, moose, bat, wolf, rodent and marsupial, had become extinct. Domestic cattle had long lost their wild progenitor, the aurochs. Tasmania had lost its talismanic thylacine, and South Africa had said goodbye to the quagga. While television might reassure the viewing public that woods, forests, plains and seas were throbbing with life, men like Julian Huxley and Peter Scott were thinking differently.

Quagga, photographed at London Zoo in 1864. It has been extinct for more than a hundred years

*

No child of my generation had any idea about species-loss. It did not dawn on me even as a young adult until I read *Silent Spring* and the world was sensitised by Greenpeace's campaign to Save the Whale. Working as an editor on *The Sunday Times*'s environment pages in the 1970s, I was impressed by the work of colleagues like Brian Jackman, who combined a lifelong, unconditional love of nature with an inextinguishable fury at what was being done to it. Week after week in the space we were generously allowed to fill, we excoriated those who saw dying or displaced animals only as unmourned sacrifices to economic growth. Economics was a one-eyed ghoul without soul or vision. Milton Friedman, the most influential economist of the late twentieth century, seemed to have given moral authority to the despoilers by declaring that the social responsibility of business was 'to increase its profits'. Simply that. To us, no philosophy had ever sounded more amoral; no full-stop more like a muffled drum. As the forests fell, and as fresh water was either poisoned or drained entirely away, so the casualties mounted and the anger grew. In those days it was unusual for newspapers to reserve space for environmental issues. The visionary editor Harold Evans made *The Sunday Times* an exception, and this allowed us to think of ourselves as pioneers. Speaking for myself, I must confess, nearly forty years on, to the sin of hubris.

In fact, the 'conservation movement' was not as new as I thought it was. For its earliest beginnings we have to go all the way back to the generation of my grandfathers. Even at the turn of the twentieth century it was obvious that George Perkins Marsh had not been crying wolf and that a philosophical divide was opening up. Until then, the idea that nature had a value in its own right was not something that had lodged in the minds of more than a few idealists, aka crackpots. It was one thing for a saint like Francis of Assisi to bind himself in brotherhood to birds and wolves, but for an ordinary mortal it looked like infirmity of mind. Man had dominion over all the beasts of the field. The Bible said so, and not even a saint could interfere with that. Thus began an ideological schism – still evident in almost any collision between man and nature – dividing those who believed wild places should remain pristine and inviolable (we may call them preservationists), and those who thought natural resources should be harvested sustainably (conservationists). The preservationist wing scored an early victory in 1872 with the world's first legally designated national park at Yellowstone, in the American states of Wyoming, Montana and Idaho. Even earlier, in 1867, the East Riding Association for the Protection of Sea Birds, whose purpose was to oppose the culling of birds off Flamborough Head in Yorkshire, had set itself up as the first-ever wildlife protection body. It was notably led by women campaigning against the harvesting of plumage for the hat trade.

Ornithologists have been the single most important group in the conservation movement ever since. I confess I am not one of them. Though I enjoy looking at birds in the garden and keep a pair of binoculars handy for the purpose, my occasional attempts at serious birdwatching have always failed through impatience, observational incompetence and intolerance of damp and cold. The failure is especially gross since the part of Norfolk in which I live – a waterscape of creeks, mudflats, sandbars and coastal marshes – is a fabled birdwatching hot spot. I'm not *entirely* hopeless. I can distinguish common-or-garden tits, finches and corvids, the commoner species of wild geese and some of the ducks, gulls and waders. My garden seethes with woodpigeons, collared doves and pheasants, and I am in thrall to the barn owl – *Titus alba*, titular parent and historic regurgitator of the Somali golden mole – which hunts across the neighbouring hayfield. But that's about it. I'm useless with anything small and brown, and (this time really to my chagrin) with almost all birds of prey save the hovering kestrel. My naturalist friends necessarily treat me like some kind of imbecile.

We pay attention to the birdmen now, but at the turn of the twentieth century they had the whole world to wake up. More than anyone, they understood the meaning of 'ecology' – a word then newly minted – and of the critical importance of habitat. George Perkins Marsh had given full and prescient warning of the consequences of deforestation, but it was the effect of hunting on African mammals that had begun to focus political minds and had united the birdmen in their cause. It is important to understand how different the world now is. During the game season, Norfolk sounds like a re-enactment of the Boer War (which in the early 1900s would have been an all-too-recent memory in southern Africa). But for all their sound and fury, the pheasant shoots are more social gatherings of like-trousered friends than serious assaults on nature or any pretence of backwoodsmanship. As the great Australian zoologist Tim Flannery has said, most of us now live in a state of civilised imbecility, less able to fend for ourselves than any man, woman or child in any society before our own. To catch, kill, paunch, skin, joint and cook an animal is as far beyond the scope of most adult males as online banking would have been to Billy the Kid. At a time when we need more than ever to mend our relationship with the world, we continue to breed generations of what Flannery calls 'poxed, inadequate weaklings', for whom a shift to self-sufficiency would be a sentence of death. Modern man can skin a client, but not a rabbit. Back in the 1890s it was different. No gentleman would have been unfamiliar with sporting guns or squeamish in their use. It was indeed English gentlemen – and English gentlemen of the highest sort – who gave impetus to the world's very first international environmental organisation, the Society for the Preservation of the Wild Fauna of the Empire (SPWFE), in 1903.

By then the damage was obvious. Erosion caused by deforestation and burning of the veldt was setting off alarms in the minds of a prescient few, but it was the staggering loss of shootable game that worried the gents. Then as now there was scepticism at the hunters' self-justifying claims that they were conservationists committed to the well-being of the very species they liked to take aim at. Nevertheless, just as we are obliged to admit that wildfowlers and game shooters in modern Britain have protected acres of habitat that would have been lost without them, so it was that the sportsmen of the early nineteenth century did rather more than just safeguard their private interests. The ivory trade had already played havoc with elephants. In the Eastern Cape, none had been seen for seventy years. In Natal, exports of ivory had collapsed from 19 tons (950 elephants' worth) in 1877 to 66 pounds in 1895. After 1880 so few elephants remained that ivory-hunters south of the Zambezi had to look for new employment. My figures are third-hand (from John McCormick's *The Global Environmental Movement*, quoting John Pringle's *The Conservationists and the Killers*), but they are easy to believe. In 1866 a single company in Orange Free State exported the skins of 152,000 blesbok and wildebeest. In 1873 it shipped out 62,000 wildebeest and zebra. With the added impacts of big-game and specimen hunting, fears of extinction – local, if not global – had begun to seem somewhat less than fanciful. The blaubuck had long gone, and so by now had the quagga. It was events in the Sudan that finally prodded the English gentlemen into action. When he learned that a well-stocked nature reserve north of the Sobat River was to be abandoned in favour of poorer territory to the south, the Verderer of Epping Forest, Edward North Buxton, gathered signatures for a letter to Sudan's Governor-General, Lord Cromer.

This was no ordinary public petition. Signatories included the Duke and Duchess of Bedford, the future foreign secretary Sir Edward Grey, Philip Lutley Sclater (for forty-two years Secretary of the Zoological Society of London), the explorer Sir Harry Johnston and his recent antagonist Professor Ray Lankester, director of the Natural History Museum, who had mocked him for believing in the okapi. Also lending his name, perhaps more significantly, was the Natural History Museum's favourite marksman, Frederick Courteney Selous. This was the group that would form the SPWFE and would develop rapidly into one of the most exclusive institutions in the English-speaking world. Its vice presidents included Lord Cromer; Lord Milner, Governor of the Cape Colony and High Commissioner for Southern Africa; Lord Curzon, Viceroy of India; and Lord Minto, Governor-General of Canada. President Theodore Roosevelt, himself a keen big-game hunter, was among the honorary members, as were Lord Kitchener and Alfred Lyttelton, Britain's Secretary of State for the Colonies.

Another important signatory was the zoologist Oldfield Thomas, who described and catalogued some 2,000 new mammals for the Natural History Museum. It might be supposed that he at least would have had some knowledge and understanding of the golden moles and other tiny basement-dwellers that lurked in the southern dark, and there were perhaps a few others whose knowledge of birds would have given them some understanding of how species related to each other. But it is difficult to imagine that many of these great men would have felt much concern for nature's lower orders. Their interest was selective. What they sought to ensure was a continuing supply of species big enough to be shot at, not the sort of creature that might turn up in an owl pellet. It earned them a sobriquet – 'penitent butchers' – and set a precedent which, even now, conservationists find hard to live down. The white man helps himself to Africa's wildlife, and blankets the earth with greenhouses gases, then tells the rest of the world it must not do the same.

The Society for the Preservation of the Wild Fauna of the Empire nevertheless deserves our gratitude. Though it saw itself somewhat disingenuously as 'a modest and unpretentious group of gentlemen', it did aim to put its influence to good use, urging the protection 'from appalling destruction' of wild animals throughout the British Empire. As the Empire at that time covered a quarter of the globe, and as the gentlemen themselves were nothing if not influential, this was no mere token. The society since has been through various name changes. After the First World War it dropped the 'Wild' to become just the Society for the Preservation of the Fauna of the Empire. In 1950 it became simply the Fauna Preservation Society, then extended its interests in 1981 to become the Fauna and Flora Preservation Society. Over time it has metamorphosed from an elite club of colonial administrators into, now, the thoroughly modern Fauna & Flora International, one of the world's most effective champions of biodiversity.

Even so, it was not until 1948, forty-five years after the foundation of the SPWFE, that George Perkins Marsh's *Man and Nature* was supplanted as the pre-eminent published authority. Again the writer was a distinguished American – Fairfield Osborn, president of the New York Zoological Society. Again the message was desperate, and again the book – *Our Plundered Planet* – was a bestseller endorsed by the intelligentsia. My own copy (a London edition) bears encomia from Eleanor Roosevelt, Aldous Huxley and Albert Einstein. 'Reading it,' said Einstein, 'one feels very keenly how futile most of our political quarrels are compared with the basic realities of life.' Well, just so. The timing was both significant and symbolic. In his introduction, Osborn explained that the inspiration for the book came towards the end of the Second World War, when it seemed to him that mankind was involved in two major conflicts. 'This other world-wide war, still continuing, is bringing more widespread distress to the human race than any that has resulted from armed conflict. It contains potentialities of ultimate disaster greater even than would follow the misuse of atomic power. This other war is man's conflict with nature.'

In recent years I have read and reviewed some nightmare visions of catastrophe. Yet none of them conjures a picture more apocalyptic than Fairfield Osborn's. 'Blind to the need of co-operating with nature,' he writes, 'man is destroying the sources of his life. Another century like the last and civilisation will be facing its final crisis.' This was, remember, 1948. Like Marsh, Osborn was a blunt-spoken polymath not much inhibited by sensitivity. His views on the need for human population control must have sounded even more shocking in 1948, when the world was still counting its dead, than they do now. 'Even after his wars, too many are left alive,' he said.

He echoed Marsh in condemning deforestation and over-exploitation of land, measuring the cost in dried-up watercourses, silted rivers, erosion and vanishing wildlife – 'as deadly ultimately as any delayed-action bomb'. In the USA, he complained, timber was being felled twice as fast as it grew. 'The story of this nation in the last century as regards the use of forests, grasslands, wildlife and water sources is the most violent and the most destructive of any written in the long history of civilisation.' Fifty-seven years before the post-Katrina storm surge that would devastate New Orleans, he saw all too plainly what must come: 'How about the valley of the greatest river of them all, the Mississippi, its bed so lifted, its waters so choked, so blocked with the wash of productive lands, that the river at flood crests runs high above the streets of New Orleans? As in historical times, the power of nature in revolt will one day overwhelm the bonds that even the most ingenious modern engineer can prepare.' Fourteen years before *Silent Spring*, he even recognised the danger of DDT.

As to mismanagement of wildlife, Osborn could find no worse an example than America's own treatment of the bison. The white man had arrived on the continent to find fifty million of them north of the Rio Grande. By 1905 only 500 remained – a loss of 99.999 per cent. Osborn's verdict on humankind is correspondingly bleak: 'The uncomfortable truth is that man during innumerable past ages has been a predator – a hunter, a meat eater and a killer.' Comparisons with other species worked only to man's disadvantage. 'His nearest relatives in the animal world most similar to him physiologically remained vegetarians. And at no time, even to the present day, have depended upon the lives of other living creatures for their own survival.'

As he saw it, the result was a mounting catalogue of devastation in which the old world fared no better than the new. A traveller in Greece told him that 'during all his travels through the mountain section of the country he saw only two pair of partridges and one rabbit – all the natural wild life having been killed off'. In North Africa, 'wandering tribes of herdsmen move from oasis to oasis, their herds stripping such grass as there is from the gullied slopes, leaving nothing but the raw unstable soil'. In southern Africa, animal life continued to be spent as if it were both a limitless resource and an offence to the sovereignty of man. 'Alarming reports have come from southern Rhodesia to the effect that more than 300,000 native wild animals have been deliberately destroyed in recent years on the grounds that they were carriers of the tsetse fly pest. This move on the part of the Rhodesian authorities, unfortunately being imitated in neighbouring territories, may well prove to be a misguided and futile butchering of the superb wild life of those regions . . . [This is] typical of man's lack of understanding of the place that wild living things occupy in the economy of nature.'

Gratifyingly to this latter-day seeker of moles, Osborn celebrated the work of burrowing animals in maintaining the health of the soil. But he had little confidence that his vision would be shared. 'It is amazing how far one has to travel to find a person, even among those most widely informed, who is aware of the processes of mounting destruction that we are inflicting upon our life sources.'

One place he might usefully have visited was the cricket-loving English county of Gloucestershire. It was here, two years earlier at Slimbridge in 1946, that Peter Scott established the Severn Wildlife Trust – the small but potent beginnings of what would grow to become the Wildlife and Wetland Trust, now one of the most effective guardians of wetland habitat, with nine reserves strung across Britain. It was indeed Britain that would become the control centre of international efforts to conserve animals and their habitats. The crucial last shove came from the recently knighted Julian Huxley in a series of three articles for the British Sunday newspaper the *Observer*, published in November 1960. Huxley had just returned from what he described as 'the most interesting assignment I have ever had' – a three-month journey through ten African countries to prepare a report for UNESCO on 'The Conservation of Wild Life and Natural Resources in Central and East Africa'. The result was a powerful mix of excitement and foreboding. His unconcealed sense of awe leapfrogs back in time over the sombre gloom of Osborn and Marsh to reawaken the ghost of Alfred Russel Wallace.

The variety of Eastern African mammals is astonishing and so are their numbers. There is still an abundance of relatively easily visible creatures — elephants, hippos, warthogs, rhinos, giraffes, lions, leopards, servals, chee-tahs, hyenas, zebras, buffaloes and baboons, monkeys and mongooses, hyraxes and hares, and a unique array of antelopes large and small – eland, hartebeest, topi, oryx, sable, roan, gnu (wildebeest), kudu, waterbuck and gerenuk, to lechwe, gazelles, bushbuck, reedbuck, impala, steinbuck and klipspringer to the little duikers and tiny dikdik.

And the sight of great herds of topi or gnu or zebra galloping across the open plains, of a troop of elephants coming down to drink and play, of a pride of lions on a kill, of sausage-like hippos in and out of the water, of a herd of impala leaping in all directions, of prehistory incar-nate in a rhinoceros, of a family of giraffes cantering along like elongated rocking horses – any of these is unforgettable, a unique contribution to the riches of our experience.

Besides these, there are many less frequently seen but wonderfully interesting mammals – chimpanzees and gorillas, bongo and situtunga, bushpigs and giant forest hogs, wild dogs and bat-eared foxes, otters and wildcats, civets and genets, polecats and honey badgers, furred mole rats and naked sand rats, elephant shrews and bush babies, porcupines and pangolins, springhares, squirrels and the strange nocturnal aardvarks and aardwolves.

Unlike Wallace he made no mention of golden moles, but one feels they were there in spirit. This was indeed the picture of Africa I had grown up with, implanted in my mind by Attenborough, Scott, and Armand and Michaela Denis. Huxley, however, was not deceived. Though there were still 'a great many' wild animals left, he noted that they were patchily distributed over a very wide area, and that their numbers were 'grievously reduced'.

A century ago South Africa harboured tens of millions of large mammals: to-day they survive in any density only in a few National Parks and Reserves. Many parts of Kenya and Tanganyika and the Rhodesias which fifty years ago were swarming with game are now bare of all large wildlife. Throughout the area, cultivation is extending, native cattle are multiplying at the expense of wild animals, poaching is becoming heavier and more organised, forests are being cut down or destroyed, means are being found to prevent cattle suffering from tsetse-borne diseases, large areas are being over-grazed and degenerating into semi-desert, and above and behind all this, the human population is inexorably mounting, to press ever harder on a limited land space.

Poaching in particular dismays him. It would have dismayed him even more if he could have imagined a time sixty years hence when park rangers – some of the bravest men on the planet – would be required to face guerrilla bands armed with rocket-launchers, AK-47s and NATO-standard Heckler & Koch G3 battle rifles. But then, as now, game departments and national park authorities had too few men to stop the poachers. Then, as now, the criminals' incentive was not just an appetite for meat.

> It is also Asian superstition and European taste for 'curios'. Indians and Chinese believe (on the basis of purely magical reasoning akin to that which led medieval herbalists to the doctrine of signatures) that rhinoceros horn is a potent aphrodisiac: they believe it so firmly that it now fetches an extremely high price per pound – much more than the best ivory; in consequence rhinos are being poached out of existence, except where well protected.
>
> Many giraffes are slaughtered by poachers merely to sell their tails for fly whisks; many Colobus monkeys to make rugs out of their lovely black and white fur; many elephants to satisfy the demands of white men for ivory ornaments, usually of low aesthetic value.

He was not entirely a pessimist, but his faith in reason would turn out to have little more foundation than the 'god hypothesis' he rejected (he was, after all, a Huxley). The route to salvation, he believed, lay in licensed game-cropping schemes that would 'go far to satisfy the Africans' legitimate meat-hunger' and so reduce the poachers' incentive to hunt illegally. It would also 'help them to realise that African wildlife is a major resource'. Somewhere at a theoretical level this truism has always been recognised. On the plains, however, the 'major resource' would be viewed in a rather narrower sense than

Huxley had intended. Inside government as well as out, private gain and common good are locked in eternal enmity. During a visit to a southern African village not long ago, protocol required my hosts to introduce me to a local official of the national government, responsible for protecting wildlife. He turned out to be a cocky, tough-looking young man accompanied by muscular sidekicks, who saw no need during our brief interview to remove his reflective sunglasses. It was like meeting the Tonton Macoute. He took little trouble to conceal the fact that he combined his official duties as nature's protector with a profitable sideline as the local Mr Big in the illegal bushmeat trade. But it is not false optimism for which Huxley's articles should be remembered. Fifty years on, it is hard to believe that a short series of pieces in a Sunday newspaper could have a lasting effect on public opinion, let alone on the care and governance of the natural world. But that is exactly what they did.

Through his director-generalship of UNESCO, Huxley had already got the International Union for the Conservation of Nature up and running. This was a huge testament to his diplomatic skill and powers of persuasion. The worldwide fellowship of naturalists and conservationists was far from the bonded coterie of like-minded fur-freaks that people tend to imagine. The international rivalries, jealousies, theological hair-splitting and clashes of personality would have done credit to any faction competing for the legacy of Marx and Lenin, or for control of the Vatican. A nest of vipers by comparison would have looked like a model of friendship. For a while it had seemed that the global centre of conservationism would drift to the USA, but the death of Franklin D. Roosevelt – a president committed to the conservationist cause – somewhat stalled the American impetus. In any case, with the focus on what was still called 'big game' in Africa, Eurocentrism had a certain political logic. At the end of the Second World War, most of Africa was still governed by European colonial powers. After a long, frustrating and often unedifying tussle for supremacy, which revolved mostly around the Swiss, Huxley used his political clout to convene an international conference to establish what at first would be the International Union for the Protection of Nature (IUPN), and what in 1956 would become, as it remains today, the International Union for the Conservation of Nature and Natural Resources (IUCN). The conference at Fontainebleau, from 30 September to 7 October 1948, involved twenty-three governments, 126 national institutions and eight international organisations. They resolved that the new body would 'collect, analyse, interpret and disseminate information about "the protection of nature"'. The mission statement has evolved with its burgeoning ambition – it now pledges to 'help the world find pragmatic solutions to our most pressing environment and development challenges' – but the dissemination of information remains at its core. Much of the information is unwelcome – the *Red List of Threatened Species* makes hard reading for anyone excited by the visions of Alfred Russel Wallace or the films of Armand and Michaela Denis. But without it the zoos, learned societies and conservation charities dedicated to the survival of wildlife would be like the blind watchmaker, struggling to put together complex life-systems in the dark.

Information was just the starting point. On its own it might help conservationists define their objectives but it didn't provide the means of carrying them out. If saving the giant panda or black rhinoceros was to be more than just a pious hope, then some means of raising money would have to be found. Once again all eyes were on Huxley. Having been instrumental in driving forward the information network, he was now the catalyst for effective action. It is a long story to which my brief account will do scant justice, but the response to Huxley's *Observer* articles was electrifying. Many others would be centrally involved – most importantly the director-general of Britain's Nature Conservancy, Max Nicholson – but it was Huxley who inspired the foundation of what is now the world's biggest non-governmental conservation organisation, WWF (originally the World Wildlife Fund, and since 1986 the Worldwide Fund for Nature). Its official launch was at the Royal Society of Arts in London on 28 September 1961, when the speakers included Peter Scott and Huxley himself. Scott became WWF's energetic first vice-president, and recruited Prince Bernhard of the Netherlands and the Duke of Edinburgh as its international and UK presidents. He also designed the WWF's famous panda logo.

Progress was rarely smooth. Conflicts with other conservation bodies, especially those in America, were par for the political course. Fairfield Osborn, who by now had founded America's Conservation Foundation, was initially a board member of WWF-US, but resigned and refused to be a trustee. (The Conservation Foundation would not merge with WWF until the 1980s, and even then WWF-US, along with Canada, would not accept the name change adopted by every other country.) Elsewhere, the royal figureheads' penchant for hunting would cause controversy reminiscent of the 'penitent butchers'. So, later, would sponsorship from oil and agro-chemical companies. There were spats with the IUCN (with which for many years it shared an office) and with the Fauna Preservation Society (over who should take the credit for saving the Arabian oryx). But gradually, over the years, like magnetised particles the forces of conservation would turn and point in the same direction.

It is a coincidence that my own idealised and wildly opti-
mistic notion of African wildlife should have developed during
that early, critical post-war period. Coincidence, too, of a
happier sort, that a couple of decades later I should find
myself sitting alongside my boyhood heroes, Attenborough
and Scott, on the judging panel of an environmental essay
competition run by *The Sunday Times* in memory of the nature
writer and broadcaster Kenneth Allsop. I knew little of Julian
Huxley then (and had certainly never read his *Observer* articles,
which were published while I was still at school), and would
never see or hear him speak – he died, aged eighty-seven, in
1975 – but his voice and philosophy, however unconsciously,
were fundamental to every entry we received. It is only now
that I realise the centrality of his thinking to everything I
believe about wildlife. How else, I wonder, would I see any
point in searching for the world's most improbable mole?

CHAPTER SIX

Resurrection

Aman at a party, name of Chris, has heard I am writing a book and wants to know what it's about. I tell him it's about an owl pellet. He laughs at what he takes to be a joke, and wishes me luck with my 'novel'. Sometimes it does feel like that. Men like Roualeyn Gordon-Cumming, Abraham Dee Bartlett, Frank Buckland, Carl Hagenbeck and Frank Buck could so easily have stepped from the pages of fiction. I am surprised, too, by the elusiveness of facts, and fall into the layman's errors of expecting simple answers to simple questions, of failing to understand the thinness of the membrane between fact and fantasy. The evidence for the Somali golden mole would hardly fill a teaspoon. Even if I do find it, I will have to rely on others to 'read' it for me. But then of course it was the very uncertainty of its existence that caught my imagination in the first place. If I wanted to be certain of success I would have searched for a baboon.

In Chapter One I mentioned some research from the University of Queensland, published in 2010, which argued that a third of all mammals previously thought extinct were actually still alive. The story flickered briefly across the news pages and then fluttered into the archive, leaving nothing behind but un-answered questions. If so many species are returning from the

dead, then why are we being warned of a looming mass extinction, the worst since the dinosaurs? And how can we be sure
that a species has completely died out, leaving not a single
individual anywhere in the world? In some cases, it seemed to
me, when the only evidence was an owl pellet, there was reason
to doubt that they had ever existed in the first place. It was like
a door opening on to a maze with an enigma in the middle and
a riddle at each corner.

At this time I had little idea of what I was getting into. I knew
only that the endless churn of species – discoveries, rediscoveries,
extinctions – was something I would like to write about. Working
on a piece for *The Sunday Times Magazine*, I asked the curator
of mammals at the Natural History Museum, Paula Jenkins, if
I could be shown specimens of some of the so-called 'Lazarus
species' stored in her collection. Thomas Henry Huxley would
tap-dance through the Central Hall if he could see how completely
his vision for the museum has displaced Richard Owen's
worshipful celebration of the Old Testament. The exhibition
space is just the smile on the museum's face. The soul, brain
and guts of the place are in the huge scientific body that stretches
out behind it. Trolley squeaking like a tumbrel, Paula Jenkins
leads me through endless back corridors, past numberless cabinets crammed with skulls, skins and skeletons. There is a vaguely
hospital-like smell which intensifies as she opens a drawer and,
from among the family *Cheirogaleidae*, lifts out the body of a
tiny animal. It lies on the trolley with its arms pinned tightly to
its sides as if it died while standing to attention. Like all museum
specimens, it is gutless, boneless and scented with insecticide.
Like many others in this faunal mausoleum, it is the holotype,
the original collected item from which the species was first
described. When Paula Jenkins turns it to the light, its babyish
face wears an expression of pained astonishment. This is the

hairy-eared dwarf lemur, *Allocebus trichotis*, which has lain in its drawer ever since a man named Crossley shot it in north-east Madagascar in 1875. Paula has worked at the museum for forty years but can't remember anyone else ever wanting to see it.

According to the late American naturalist Francis Harper, not a single living example of *Allocebus trichotis* was seen between 1875 and 1945, when Harper himself declared it extinct. This was not an arbitrary personal opinion but the straightforward application of a widely accepted scientific principle. It stood to reason, it was a *fact*, that any creature not seen for fifty years must have disappeared for ever. That was how extinction was defined. Goodbye, dodo. Goodbye, *Allocebus trichotis*. End of story. Then in 1967 came the miracle. A researcher in Madagascar reached into a hole in a tree and out came the hairy-eared dwarf lemur. It would prove, however, to be the very briefest of resurrections. Darkness closed again, and there was not another 'official' sighting until 1989, when WWF found it near the Mananara River. Crucially this time they took the trouble to interview the locals, to whom *Allocebus trichotis* was anything but a mystery. The bare facts of its existence were as follows:

It was a very small animal – head and body typically between 125 and 145 millimetres, tail between 150 and 195 millimetres, weight between 75 and 98 grams. It was nocturnal and nested in forest trees, usually between three and five metres above the ground. Throughout the cold season, May to mid-October, it hibernated deep inside tree holes. All this, of course, helped to explain its supposed 'extinction' and to prove the inadequacy of the fifty-year rule. If you wanted to see *Allocebus trichotis*, then you would have to go and look for it. It wasn't going to scuttle out of the forest and say, 'Hi, I'm not extinct.' Even the Madagascans saw it only between October and March, which is the tree-cutting season in the annual cycle of slash-and-burn.

The next specimen on Paula Jenkins's trolley is something
that looks like an inflated shrew. Unusually, it is stuffed and
mounted as if for public display – a vulgar, unscientific practice
abandoned by the museum in the late nineteenth century (this
one was collected in 1898). The Cuban solenodon, *Solenodon
cubanus*, is from the outer fringes of the mammalian estate,
where weirdness lies. It is a genuine primitive, similar in many
ways to the early mammals that followed the dinosaurs. The
long bendy snout looks frankly comical – a clown's proboscis
– though the naked, rat-like tail is less endearing. Less attractive
still are its fangs, through which it can inject venom like a snake.
You wouldn't find Michaela Denis cuddling one of these, but
then she almost certainly never saw one. By 1970, after none
had been seen since 1890, the Cuban solenodon was officially
declared extinct. It remained non-existent for just four years
before confounding its obituary and reappearing in Cuba's
Oriente Province. Having no faith in apparitions or ghosts,
science had to admit its error. *Solenodon cubanus* might be endangered, but it was definitely not extinct.

The Natural History Museum has many other examples.
Jentink's duiker, whose skull is next on the trolley, was not seen
alive between 1889 and 1948. The Cyprus spiny mouse had been
lost for twenty-seven years before it popped up again in 2007.
Fea's muntjac went missing between 1914 and 1977. The thylacine
has not been seen in the wild since 1933, though there are plenty
of amateur enthusiasts in its native Tasmania (by no means all of
them obvious hoaxers or wishful romantics) who are prepared to
say otherwise. Reported sightings, including photographs and at
least one video clip, comfortably outnumber glimpses of the Loch
Ness Monster. More than all the other specimens I see on this
visit, the two stiffened pelts Paula Jenkins now lays on the tumbrel
bear a colossal weight of tragedy. The thylacine, popularly known

as the Tasmanian tiger, was a gloriously improbable assemblage of unmatched parts whose scientific name *Thylacinus cynocephalus* ('dog-headed pouched one') precisely describes its uniqueness. It was a carnivorous, dog-like marsupial with exceptionally wide-opening, bone-crunching jaws and a conspicuously striped rump. It was likely to have been near-extinct on the Australian mainland even when the museum's two specimens were collected in 1839 and 1865, but it managed to cling on in Tasmania. Even there, however, its taste for sheep made it anathema to farmers, who shot it on sight, and it would have taken a determined conservation effort to save it. Tragically, no such effort was made. The last scientifically validated individual was caught in 1933 and died in Hobart Zoo three years later. Since then – nothing. More than twenty-five scientific expeditions have done their best to confirm the amateur sightings, but all have ended in failure. The thylacine was declared extinct in 1982.

Thylacine – the last known survivor died in 1936, but enthusiasts refuse to believe it is extinct. This one was photographed at London Zoo in 1913

Next we head for the *bovidae*, the family of sheep, goats, cattle, antelopes and buffalo. Doubts about provenance are already raising questions about the very existence of some species – a line of thought that will lead directly to the golden mole, though it's antelopes that come to my attention first. I am still a stranger to the idea that entire species might be open to doubt; that the evidence for their existence might be too flimsy to pass forensic scrutiny. The red gazelle, *Eudorcas rufina*, for example, has never been seen alive. Its entire classification depends on three mysterious specimens bought at Algerian markets in the late nineteenth century. According to the IUCN, 'most authors' have accepted it as a genuine species, though 'continuing doubt concerning the validity of this taxon' has persuaded it to change the classification from 'Extinct' to the more enigmatic 'Data deficient'. None of the three specimens is in the Natural History Museum, but I have been expecting to see a similar rarity, the Arabian gazelle, *Gazella arabica*. Sadly there has been a crossing of wires. The evidence for *G. arabica* is even thinner than it is for *E. rufina*, and I am mistaken in my belief that I will find it in Kensington.

In fact, the only known example of the Arabian gazelle is a single male specimen in Berlin. It was apparently collected in 1825, when it was reported somewhat dubiously to have come from the Farasan Islands in the Red Sea. Some scientists have suggested that the *Red List* classification of 'Extinct' should be changed, like the Red gazelle's, to 'Data deficient'. It is, they argue, illogical to classify as extinct a species which, in their opinion, never existed in the first place. Disappointingly I have muddled it with *Gazella bilkis*, which some authorities have classified as a subspecies of *G. arabica*. My mind whirls. A *subspecies* of a species that may never have existed? How am I supposed to make sense of that? Paula Jenkins tells me that the museum's specimen is one of only five known to exist, all of which were collected in 1951 from Yemen – hence its English name, Yemen gazelle, or, more poetically, Queen of Sheba's gazelle. Arguments about its status now are literally academic. 'There is no doubt,' says the *Red List*, 'that the population originally described as *G. bilkis* is certainly now extinct, regardless of whether it was a species or a subspecies.'

Moving from cabinet to cabinet, I am slowly adjusting to the idea that *Mammal Species of the World*, the *Red List* and all other attempts to gazette the world's fauna are best guesses rather than audited accounts. Estimates of the total numbers of plant and animal species vary so widely that they look less like science than a soothsayer's reading of entrails. Guesses roam between two million and 100 million, though most of them fall between five million and thirty million. Of these, only about 1.8 million have been described and catalogued, and only 3 per cent of these 'known' species have the benefit of IUCN status reports. Fifteen per cent of mammals are classified as 'Data deficient', meaning that we have very little idea of their range, number or chance of survival.

Complicating matters further, 'new' species continue to turn up. Only a few weeks earlier, a previously unknown mammal had been named and announced to science and the world's press. *New Carnivore Discovered in Madagascar*, said the headlines, over pictures of something variously described as 'mongoose-like' or 'a scruffy ferret'. The holotype, consisting of empty skin, bare skull and mandible, now lies in a box in Paula Jenkins's office. It is small – only slightly larger than the hairy-eared dwarf lemur – but in death, with its fluffy tail and grizzled pelt, it looks rather sleeker than it did in life, a masterpiece of the mortician's art.

This is *Salanoia durrelli*, or Durrell's vontsira, named for the eponymous Gerald, late naturalist, founder of Jersey Zoo and author of *My Family and Other Animals*. It comes from the wetlands of central eastern Madagascar, where it feeds on small mammals and fish. Only three have ever been identified, which – taking account of the limp skin now lying in front of me – leaves exactly two recorded in the wild. The unavoidable suspicion is that its discovery has only narrowly preceded its extinction. Like the dwarf lemur, which vanished again in 1989, it is one of the most endangered species on earth.

Even quite large animals can live beyond the reach of science. It was only in 1992, for example, that the saola, the uniquely beautiful antelope *Pseudoryx nghetinhensis*, was discovered among the Annamite Mountains of Laos and Vietnam. Then in 2010, in northern Myanmar, a team of international scientists happened across the previously unknown Myanmar snub-nosed monkey, *Rhinopithecus strykeri*. Let's not forget that the gorilla was first described only in 1848, and the okapi not until 1901 – extraordinary given their size, distinctive appearance and the intensity with which Africa had been searched. It is this very uncertainty that keeps alive the hopes of the thylacine-hunters and the indefatigable friends of Nessie. Their optimism holds a strange echo of religious faith – just because you can't see something, it doesn't mean it isn't there. Or, as scientists prefer to put it, absence of proof is not proof of absence.

Of course animals are 'new' only in the sense that America was new when Christopher Columbus bumped into it. Local people in Myanmar were as surprised by the scientists' ignorance of the snub-nosed monkey as the scientists were by its discovery. The peculiar nasal arrangement that gives the species its name makes the animals easy to find in the rain – water trickling into their upturned nostrils makes them sneeze. The one predictable thing about evolution is that it is very, very slow. New species do not pop up like mushrooms. Nor do they last for ever. Mammals typically survive for around a million years, though some may hang on for as much as ten times longer. Vague though this may be, it provides the basis for a bit of simple arithmetic. Given that there are around 5,000 known mammal species in the world, the extinction rate on average should be around one every 200 years. Over the last four centuries that rate has been exceeded by a factor of nearly forty-five. Even if a third of those missing did walk back out of the jungle, it would still be a catastrophic rate of loss.

Confusingly, too, 'new' species sometimes are just old ones in a different guise. Science is always revising and reappraising itself. Re-examination with new techniques may show that what was once thought to be a single species is actually several different ones. Thus the number of species increases, but the number of animals remains the same. This 'lumping and split-ting' has been most dramatic in birds, where the list of species since the 1970s has soared from 8,600 to around 10,000. Albeit on a smaller scale, it has happened with mammals too. The echidna is a good example. Truly this is one of nature's strangest, a spiny, tube-nosed native of Australia and New Guinea, which, with the single exception of the duck-billed platypus, is the only mammal known to lay eggs. For years it was thought that there were just two species – short-beaked and long-beaked. Then scientists began to argue. Some reckoned that the long-beaked was not one species but six. Eventually, in 1998, they settled on three. All now are listed by the IUCN as critically endangered. One of them – Sir David's long-beaked echidna, *Zaglossus atten-boroughi*, named after the world's favourite naturalist – must count as one of the rarest creatures on earth and, as we shall see later, has recently achieved a somewhat ambiguous celebrity of its own. The entire species is known from just one individual seen in 1961. But even this disqualifies it as the rarest ever recorded. At least it has been seen alive. The Somali golden mole exists only as a fragment in an owl pellet.

Bending over the tiny skull of Durrell's vontsira, I realise that questions of existence and identity are far more complex than I have understood. The old-fashioned fifty-year extinction test – not finally abandoned until 1995 – was patently absurd. The Somali golden mole – any number of rare or uncatalogued small animals – might inhabit the thorny tangles around my garden, and I would have no idea they were there. I have never systematically looked. Nor, I would guess, has any previous occupant of this land, reaching all the way back to the Neolithic. Extend that thought to all the thorny tangles, remote forests, hidden valleys, plains and mountains of the world, and one can see why science is so imprecise. It is like one of Donald Rumsfeld's 'known unknowns'. We know that in scattered populations or discrete enclaves in every continent, old species must continue to outlive their apparent deaths and 'new' ones remain undiscovered. We don't know what or where, and we don't have the resources to find out.

Back at home I return to the document that first pricked my interest – 'Correlates of rediscovery and the detectability of extinction in mammals' (authors: Diana O. Fisher and Simon P. Blomberg of the School of Biological Sciences, University of Queensland). It tells me that 70 per cent of 'purportedly extinct' mammals are known from fewer than five historic sightings. This is why the IUCN's extinction criteria now rest on search effort rather than time-lag. Now a species cannot be written off until there have been 'exhaustive surveys in known and/or expected habitat, at appropriate times, throughout its historic range . . .'

Resources for this kind of thing are limited, and there is a bias towards large iconic species that engage the public interest. By 2010, as we have seen, at least twenty-five qualified search teams and many more amateurs had mounted expeditions in search of the thylacine. The wild horse, *Equus ferus*, had been sought twenty-two times since 1969; the kouprey, or grey ox, *Bos sauveli*, twenty-three times since 1986; and the baiji, or Yangtze River dolphin, *Lipotes vexillifer*, fifteen times in four years. As the authors point out, there is a serious risk of blind faith, of searches continuing long after the quarry is extinct.

The great majority of missing mammals have been accorded little or no search effort at all. *Calcochloris tytonis* even now might be tunnelling away in Somalia, or it might not. Who could know? Somalia of course is a special case, a rogue state which is no place for rambling zoologists. But it is not alone in presenting difficulties. There are plenty of governments whose commitment to freedom does not stretch as far as inviting foreign scientists (aka 'spies') to explore their territories, and plenty of places where political obduracy is combined with remote and difficult terrain. It all militates against the small, the far-flung and the obscure – which of course is exactly what many 'Extinct', 'Critically endangered' or 'Data deficient' animals are likely to be. It will be a while before anyone invites us to adopt a pygmy spotted skunk, a Sulawesi warty pig or a Sundaic arboreal niviventer. As the conservation charities very well know, public interest is a powerful arbiter, and the public likes big charismatic mammals it can easily recognise. This is why the likes of thylacine, wild horse and grey ox have been so ardently pursued, and why public appeals are slanted towards megafauna. It is an obvious truth, too, that a big animal with a limited range is a lot easier

to spot than a small one ranging widely, and that no amount
of effort will reveal something that isn't there. On the
Queensland evidence, it is search No. 12 that marks the point
at which persistence turns to folly. No lost species sought
more than eleven times has ever been found.

There is another complicating factor, too. Research interest
and public support have been heavily concentrated on animals
hit by persecution or exploitation. This is how the international
conservation movement began, and it is how many people still
perceive it. More than twice as many searches are mounted for
animals that have been shot as for those that have lost their
habitats or been displaced by alien species. Perhaps the guilt is
sharper, the issue more emotive, but it's a mistake to imagine
that the lanyard of a chainsaw is any less lethal than the trigger
of a gun. Lingering, attritional deaths may not be as dramatic
but the animals in the end are just as dead. On the other hand,
as animals that range very widely are harder to exterminate by
gunfire than those whose ranges are small, it follows that roaming
species reduced by habitat-loss are much more likely to be
rediscovered than victims of the gun. The good news, according
to Queensland, is that the number of species thought to have
been eliminated by loss of habitat 'is likely to be overestimated'.
But of course the truth of this can't be tested without a huge
global research effort, and huge global research efforts are few
and far between.

I think again of 'my' long-lost mole. Somalia, formed in 1960
by merging a former Italian colony with a British protectorate,
was still a young country when Alberto Simonetta found his
owl pellet there in 1964. It developed rapidly into one of the
most chaotic and violent countries in the world, literally ungov-
ernable. There has been no effective central government since
the overthrow of the socialist President Siad Barre in 1991.

Tribal, political and religious factions have been at war ever since, at the cost of at least a million lives and a persistent headache for neighbouring Ethiopia and Kenya. Somali poachers are notorious plunderers of Kenyan wildlife, especially elephants, and pirates have put coastal waters off limits to any sailor without either a death wish or a naval escort. The British Foreign Office warns against all travel to Somalia, and advises visitors to Kenya not to venture within 60 kilometres of the border. Would I – would *anyone* – go there in search of a mole, even one as rare as *Calcochloris tytonis*?

Speaking for myself, the answer is no. Risks have to be in proportion to the likely gain. If I am to search anywhere, it will have to be well inside the Kenyan border, or in Florence, where Alberto Simonetta took his now apparently lost specimen in 1964. It is too much to expect that anyone in Kenya will be able to find a living example of *Calcochloris tytonis*, but the country has a very long relationship with golden moles. Fossil remains of a bygone species known as *Prochrysochloris miocaenus* date back to the Miocene epoch, between 23 and 5.3 million years ago. These days Kenya is known to harbour Stuhlmann's golden mole, *Chrysochloris stuhlmanni*, which lives also in Burundi, Cameroon, the Democratic Republic of Congo, Uganda, Rwanda and Tanzania. Its IUCN category is 'Least concern', which makes it considerably more common than some of the other animals I hope to see. Golden moles are usually described as a family of 'ancient' species which are distinct from 'true' moles, though they look and behave very like them. They have the same burrowing habit and powerful claws for digging, and spend most of their lives under ground. They are blind, and use their ears to locate the small insects and worms that are their preferred food. They conserve energy in cold weather by going into a torpor,

and have such efficient kidneys that most of them do not need
to drink. These extreme specialisations seem to argue against
the idea, put forward by some, that they are undeveloped
primitives, but the absence of a scrotum in the males, and
possession of a cloaca – a single orifice through which they
pass both urine and faeces, like a bird – are not exactly marks
of sophistication. Pictures of Stuhlmann's golden mole show
a densely furred, eyeless and iron-clawed creature with a long
sleek body like a swimmer's (desert species are indeed
described as 'swimming' through the sand). Kenya might not
offer *C. tytonis*, but to see one of these cousins would be a
major consolation. As yet I don't know how likely it might be
but, even with my optimism still undimmed, I have to reckon
it's odds against. Huge landscape, tiny subterranean animal.
Who am I kidding?

Which of course leaves the riddle of Professor Simonetta,
the Florence Institute of Zoology, and the animal unluckily
named after a failed state. Is Simonetta still alive? As he
published his paper on the Somali golden mole in 1968, it's
clearly possible. I search for him on Google and find what
appears to be a short biography written in Italian. From this,
though I can't understand much else, I gather that he was
born on 26 March 1930 and so would be in his early eighties.
I gather also that his wife died in 1999, but there is no date
for his own demise. It's a good start, which becomes better
still when I scan the document again and spot the key word,
'Somalia'. Next I find an undated paper written by him in
English, promisingly titled 'Control of poaching and the
market for products such as ivory, rhino horn, tiger and bear
body products.' It identifies him as Professor of Zoology at
the University of Florence. Looking further, I find that this
is a chapter from a book, *Biodiversity conservation and habitat*

management (Vol II), published in 2008. I might not get to read it – Amazon is asking £146 for the paperback – but it's an encouraging sign. All it should take is a call to the Università degli Studi di Firenze . . .

Then I remember. I am not the first to embark on this trail. One of the highly qualified and experienced authors of *Mammal Species of the World* – the golden mole expert Gary Bronner from the University of Cape Town – has tracked *Calcochloris tytonis* to Florence but failed to find it. Why should I – unqualified, inexperienced and no kind of expert – hope to do any better? And there is another reason why my finger hesitates over the telephone keypad. *It's just too soon.* I'm not ready for a definitive answer. Suppose I get through to Simonetta and he tells me the specimen is lost. What then? Or suppose, against all expectation, he's got it in his desk drawer. Either way, the search would be over. It's over, too, if Simonetta is no longer alive. I write down the university's number but do not dial it. *I want an excuse to go on searching.*

It's deep midwinter now. Florence, I decide, can wait for the European spring or early summer. On a bright and unseasonably warm day in the second week of January, I take a train to Cambridge. The hunt has brought me here once before, to pick the brain of Craig Hilton-Taylor, the amiable South African biologist who heads up the IUCN species programme. It was thanks to him that I knew what to look for in the Natural History Museum. Thanks to him, too, that I started browsing zoology textbooks and stumbled across the golden mole. It was also in Cambridge, during a brief fellow-commonership at Corpus Christi, that I learned the habit of not-always-disciplined research. In Jaroslav Hasek's satirical masterpiece *The Good Soldier Schweik*, a character called Cadet Biegler is said to pursue knowledge with the zeal of an idiot. Cambridge always brings out my inner Biegler. I once went to the University Library to read about men's hairstyles in the seventeenth century, and spent the entire afternoon learning about bearded women (leading world authority: the Surgeon General of the US Army). Where else would I have begun to read, species by species, through a database of mammalian taxonomy? Today, arriving early, I allow myself a short, unscheduled visit to the Fitzwilliam Museum in Trumpington Street, where I have just enough time to sprint through the Italian Renaissance before a lunch-date further up the street.

My host is Mark Rose, long-serving chief executive of Fauna & Flora International. Like most conservationists of my acquaintance, he is no prissy vegetarian. Having steered me towards the wild duck (which jogs our memories of Peter Scott), he opts for steak and a serious red wine of the kind that would have pleased the clubbable gents who founded the Society for the Preservation of the Wild Fauna of the Empire in 1903. FFI still enjoys big-name backing, but the names these days are more likely to be from the celluloid aristocracy than from the blood-lines of English nobility. Its vice-presidents include Sir David Attenborough, Dame Judi Dench, Stephen Fry and the Australian comedian Rove McManus. Cate Blanchett has also turned out in support. I wonder if celebrity endorsement really works; whether the attachment of star names doesn't actually trivialise rather than add weight to a campaign? I reflect that I long resisted the purchase of a perfectly good coffee-making machine simply because it was endorsed by a Hollywood film star. Mark is adamant that it works, provided the names are from the cerebral end of the celebrity spectrum and not reality-television airheads. Judged by this criterion, his list looks impeccable. As we have seen, the big-name tendency in wildlife conservation extends also to headline species. FFI in Africa is focusing on, among others, the lion and African wild dog in Mozambique; the black and northern white rhinoceroses in Kenya; the Pemba flying fox on Pemba Island off Tanzania; the pygmy hippopotamus in Liberia; the Cross River gorilla and western mountain gorilla in Cameroon: the eastern lowland gorilla in the Democratic Republic of Congo, and mountain gorilla in the DRC, Uganda and Rwanda. Elsewhere it is working with the Asian elephant, the Bornean orang-utan, the jaguar, the Iberian lynx, the red panda, the snow leopard, the Sumatran tiger, the Hainan gibbon (the world's rarest ape, with only twenty still surviving), the Tonkin snub-nosed monkey (which was believed extinct until it was rediscovered in the early 1990s), and the newly identified Myanmar snub-nosed monkey.

The argument for charismatic species is pretty much the same as it is for charismatic vice-presidents. They attract attention. In conservation terms the justification is that what's good for a headliner is good for every other creature that shares its territory. Habitat is for one and all, and to conserve rhino and gorilla is to conserve the golden mole. The question is: where in Africa (and for me it *has* to be Africa) should I go to see conservation at the cutting edge? Mark suggests northern Mozambique, where the huge Niassa National Reserve holds a large population of rare hunting dogs as well as a lions, leopards, elephants and spotted hyenas. It is the largest protected area in Mozambique, and one of the biggest in Africa. He is also keen on South Sudan, the brand-new country that declared its independence in July 2011 after twenty-two years of civil war had killed at least 1.5 million people and displaced millions more. FFI is now working with the national government to establish an effective conservation policy, fight the poachers and rehabilitate the ravaged but obstinately surviving wildlife (there have been rumoured sightings even of the critically endangered northern white rhino). Here are all the perils, pitfalls and pleasures of Africa in a single spectacular nutshell. There is a nice coincidence too, in that it was the threatened relocation of a nature reserve in Sudan that first brought Curzon, Kitchener, Roosevelt and the other 'penitent butchers' rushing to the aid of animals in 1903.

However, it is not to Mozambique or South Sudan, or even to the Congo, that my imagination has transported me. Two things attract me to Kenya – or three, if I count the fact that I've never been there. Both in their way are historical. Brumas apart, the animals that most excited me on childhood visits to Regent's Park or Whipsnade were all natives of Africa, and (though I may be wrong about this) I remember Armand and Michaela Denis's wildlife films being overwhelmingly a homage to Kenya. Then of course there's all the Happy Valley, *White Mischief* stuff, and Karen Blixen's *Out of Africa*. I may be up to five reasons now, but there is an even more important one to come. Julian Huxley wrote of 'prehistory incarnate in a rhinoceros'. No animal better encapsulates the awesome strangeness of Africa, its ancient and mesmerising power, than the rhino. And no animal more starkly exemplifies the desperate fight for life in which so much of wild Africa now finds itself locked. In the 1970s and '80s, poachers reduced the overall number of black rhinos from 100,000 to 4,000. The eastern subspecies is now down to 700. But this is nothing compared to the plight of the northern white species, of which (discounting unconfirmed Sudanese rumours) only four are known to exist in the wild and four more in zoos. Eighty per cent of the eastern blacks are in Kenya, and the largest single concentration of them lives within the protected area around the Ol Pejeta

Conservancy in the Laikipia district north-west of Nairobi. All four northern whites are there too. And then there is Stuhlmann's golden mole . . .

Where am I going to go? Mark Rose nods and raises his glass. It is settled.

CHAPTER SEVEN

Chopsticks

A man of strong religious conviction once wrote to me in fury – mauve ink, capital letters, heavy underlining – condemning my use of the word 'sophisticated' as a term of approbation. For me, sophistication had meant refinement (in this particular case, the subtle interpretation of a complex argument about animal rights). For him it meant something not far removed from blasphemy, the exact opposite of the artless simplicity with which he framed his prayers. To him, sophistication was the enemy of innocence, and hence of Christian integrity. I didn't agree, but it made me think.

There are many things I regret about growing old, but first among them is the loss of innocence. I mean this in a different sense to my fulminating correspondent. It is not that experience has corrupted me. On the contrary. I entered the world as a screaming savage, and it is experience that has moulded me into a more or less tolerable member of society. I mean only that age has dimmed my vision. Nostalgia is not homesickness, nor any misplaced craving for a Golden Age that never existed. What I miss is childhood's eager eye, the capacity to look at the world and be amazed. My first mind-altering experience with Dartmoor is unrepeatable. I can return to the spot – I do it often, and always love what I see – but it's the same rabbit from the same hat. It's wonderful but it's not magic. Excitement is dulled by repetition, expectation fulfilled but not transcended. For me, the pleasure of travel is in rediscovering that elemental way of looking, the joy of never-before. Into my own mental storehouse, never to be forgotten, went the first, garlic-and-Gauloises whiff of France; the first view of the ground from an aircraft; the first shock of Mediterranean heat; the first ride in a car at over 60mph (an exhilarating speed in the 1950s); my first unaided swim. Later would come the first glimpses of Versailles, Venice, Botticelli's *Venus*, the Alps, Marrakesh, an Icelandic glacier, a humpback whale.

Soon will come the first wild rhinoceros I've ever seen, and the first lions and elephants outside circuses and zoos. If I have given the impression that I am some kind of old Africa hand, then let me now dispel it. Discounting Egypt and Morocco, I have been to the continent just three times, and each time to the same country, Mozambique. These were big experiences, but not the kind that reawaken the sleeping child. As a journalist I had gone to record a blighted country's loss, and its attempted recovery, after sixteen years of civil war.

Almost exactly in the middle of Mozambique, at the southern end of the Great African Rift Valley, midway between Zimbabwe and the Indian Ocean, lies Gorongosa National Park. Before the war its 4,000 square kilometres of forest and savannah was one of the glories of Africa. Its stylish headquarters at Chitengo Camp was (sorry, mauve-ink man) a sophisticated retreat for the fashionably rich, who could enjoy the sound of lions over their cocktails. War changed everything. On my initial visit in 2005 the first thing I noticed was a red rag hanging from a stick. Beneath it, poking through the dust, were two unexploded mortars. A few yards away, in the roofless shell of a bombed-out schoolroom, two men squatted by a fire. Through an interpreter

I learned that they would be here for three months, working off a fine they couldn't pay for poaching warthog. What had once been a resort was now an open prison.

During the civil war, Gorongosa was the heartland of the Renamo guerrillas, for whom trees were fuel and wildlife was meat. Chitengo was blown to bits, its elegant bars, restaurant and pavilions mortared from within, its swimming pool reduced to a shallow, slime-green sump. A bare coiled spring was all that remained of the diving board, and not much more was left of Gorongosa's wildlife. Numbers of elephants during the war shrank from 4,500 to 200, hippos from 4,000 to 62, lions from 300 to 25, zebras from 20,000 to 60, wildebeest from 20,000 to 50, and so on, all the way down to soil invertebrates. In a year there had been no sign of leopard or cheetah, and plains that should have been swarming with antelope and wildebeest were rolling oceans of head-high grass. The only animals in any kind of abundance were warthogs and baboons, which people in the villages surreptitiously killed and ate. Twelve years into the peace, the despoliation had yet to stop. A tiled bathroom in one of the old safari lodges contained a rusty arsenal of weapons confiscated from poachers. Heaped against the wall were machetes, knives, bows and arrows tipped with hammered barbed wire or sharpened strips cut from old car doors; buffalo-size snares; gin traps made from vehicle springs; 200-year-old cap-lock rifles complete with wadding, home-made gunpowder and misshapen hand-made bullets. This was Mozambican roulette. A gun like this may fire when you pulled the trigger; or it might explode and blow your head off. Such are the economics of desperation.

On a drive through the park I saw a bushbuck, a few gazelles, a crocodile and some trees uprooted by an elephant. I was told that lions had returned but I neither saw nor heard any. During the night and early morning, the only sounds were birds and the drilling of novice park rangers, dressed in rags and presenting arms with sticks. Many of them, I was told, lived by poaching.

When I went back in 2009, the American philanthropist Greg Carr, working with the Mozambican government, had begun the long-term restoration of the park. Life there was still hard. In the shed-sized medical centre I found a shirtless young boy, apparently in a catatonic trance, being prodded by a nurse. He looked no older than twelve but his father said he was sixteen. He had had malaria and now had pains in his stomach. The 'ambulance' for the three-hour lurch to hospital in Beira would be a filthy Nissan pick-up with a mattress in the back. His father obligingly spelled out the boy's name but showed no sign of distress. This is just the way it is. A few moments later a park ranger approached the Portuguese manager and asked for time off to bury his baby.

Since then Chitengo has been substantially rebuilt and once again is open for business. There are luxurious thatched cabins for tourists to sleep in; a new swimming pool, gift shop, restaurant, morning and evening game drives, safari trails and that new essential for survival in the bush, Internet connectivity. The park itself is gradually being brought back to life. Zebra, wildebeest and buffalo have been reintroduced to graze the plains, which they share with elephant, oribi, reedbuck, waterbuck, warthog, sable, impala and lion. It is an odd but interesting reversal of polarities. Once the wildlife brought in the tourists; now tourism brings in the wildlife.

Other things do not change, or if they do it's for the worse. Late on a November morning in that same year, 2009, a band of men strolled up to a ramshackle farm on the Kenyan side of the Tanzanian border. Inside, the farmer was sharing a beer with a couple of friends. Outside, a pot of vegetables steamed over a fire, watched by his wife. Children swarmed after a punctured football; hens squabbled in the mud. At first, the appearance of strangers aroused only curiosity. This was poor country where meagre livings were scraped from exhausted soil, but it was peaceful and not unwelcoming. Then the farmer saw the guns.

The day was wet, the beginning of the short rains, a day that many would never forget. The farmer and his friends did not know it, but theirs was just one small incident in a war growing hotter by the day. For dozens of farmers and shopkeepers spread across six countries, it was a morning of unwelcome shocks. All the Kenyans knew was that they were in big trouble. Outnumbered and outgunned, they had no choice but to let the raiders take what they wanted. This war was not for land. It was not for oil or diamonds or gold. It was, and still is, for an essentially useless commodity over which men fight as if for their lives. *Ivory*. The command centres in this war are far from the killing fields. They are in Beijing, Tokyo, Hong Kong and Bangkok, not in the threadbare *shambas* of sub-Saharan Africa. It is a foot-soldier's war being waged with the very best tools the arms industry can provide. Rocket-launchers and AK-47s do their job well, and the world is rewarded with chopsticks.

The men at the farm were lucky. Their visitors were rangers from the Kenya Wildlife Service (KWS), part of a coordinated anti-poaching operation run by Interpol. The farmer and his friends were suspected of poaching, but their rights would be respected and they would arrive in court with their limbs intact. Such courtesies are not often reciprocated. Poaching gangs are trained to military standard and armed with automatic weapons. Elsewhere in Africa, rangers often have only obsolete Second World War rifles, a hopeless mismatch not much better than pitchforks.

At an airfield soon after their arrest, the prisoners were photo-graphed with a tusk found in the farmer's house. It was still bloody, evidently from an animal only recently dead. Locally its value was around £250. In Japan or China it would have fetched forty times as much. It's a tragedy as old as the bullet. Roualeyn Gordon-Cumming was an heroic slaughterer who perpetuated the cycle of violence by offering muskets in return for ivory. His mortification at failing to find eight or ten 'first-rate bulls' which he knew he had 'mortally wounded' was not for the animals' suffering. It was for the loss of £200-worth of ivory. He found it vexing 'to think that many, if not all of them, were lying rotting in the surrounding forest'. Another day, when tusks were 'stolen' from an elephant he had killed, he galloped to the nearest village and offered to shoot the chief. Those days of petty opportunism are as distant from twenty-first-century organised crime as Fagin's young pickpockets from the *Cosa Nostra*. Small-scale local rackets have been transformed into well-organised businesses run by international syndicates, and there is enough money in it to warp the politics of half a continent.

The ivory trade has been illegal since 1989, when it was banned by the Convention on International Trade in Endangered Species (CITES), after the worst decade for elephants in history. At least 700,000 had been killed, and photographs of their mutilated carcasses provoked revulsion wherever they were published. For a while it looked as if the ban might work. Ivory lost 90 per cent of its value, the trade dried up and poaching declined dramatically. Conservationists began to congratulate themselves on the most successful piece of international wildlife legislation ever enacted. The elephant had been saved!

But it hadn't. Some countries simply could not afford to protect their herds, and not all of them wanted a trade ban anyway. Encouraged by the efforts of Zimbabwe and some other countries to subvert it, the poachers began to sneak back in. There was nothing to stop them. Much of rural Africa is more than two days' trek from a police station. Without an army and blanket air cover, protecting a national park like Kenya's Tsavo, which is the size of Israel, is not easy. Illegal trade routes quickly reopened and the price of ivory rocketed. By 2004 it was back up to 200 dollars a kilo. Three years later it had more than quadrupled, to 850 dollars. By 2009 it was 1,200, and by late 2011 it had topped 1,400 dollars (£900). And that's just wholesale. Retail in China, you could be talking about a multiple of five or even more. One law enforcement officer saw a carved tusk offered in a Tokyo market for 250,000 dollars. Against this, human life is cheap. By 2010 the trafficking, and the killing that goes with it, were almost back to where they had been in the dark days of the 1980s.

Counting dead elephants is not an exact science. Carcasses may be covered with branches so they can't be seen from the air, and smuggled goods leave no paper trail. The best estimate is that Africa is losing 8 per cent of its elephants every year – at least 38,000 in 2009; maybe 36,000 in 2011. This sounds bad even before you understand that, even in the best of times, the animals' rate of reproduction is just 5 per cent.

The killings are brutal. In the nineteenth century, adult elephants would be shot to enable the capture of their calves. Now it happens the other way around. The calves themselves are shot – not for their ivory, for they have none, but as bait for their mothers, who will be picked off when they come to grieve. I never heard what happened to the men from the farm, but they were scarcely big enough to be called even small fry. For possessing ivory they might have faced a fine of maybe 8,000 Kenyan shillings (approximately £64). For a firearms offence they might have seen the inside of one of Kenya's notoriously unpleasant jails. But their removal from the action would have been of small benefit to the elephants, and of no consequence to the Mr Bigs. Even if it were being done legally, the industrial harvesting of six-ton animals would be far beyond the resources of petty thieves. You need infrastructure, logistics, technology, a skilled workforce, management systems and a solid client base. You also need money sufficient to buy the cooperation of customs and port officers, park managers, civil servants and ministers.

'Operation Costa', as the 2009 Interpol raids were called, did not touch a hair of the godfathers' heads. Aimed at small-time poachers and dealers in Burundi, Ethiopia, Rwanda, Tanzania, Uganda and Kenya, it turned up 1,768 kilos of tusks and carved ivory. As a mature elephant carries around eight kilos of tusk, that represents at least 220 dead animals and almost certainly more – a mountain of carnage, but still only the barrel-scrapings of a trade responsible for the deaths of 38,000 elephants a year. There were two factors, however, that hinted at the scale of the criminal hinterland. Most of the worked ivory was in the form of signature seals, cigarette holders and chopsticks, obviously intended for export. Even more worrying was the haul of fire-arms, which included German-made Heckler & Koch G3 battle rifles. These are fully automatic military weapons firing NATO standard 7.62 x 51 millimetre bullets, which can empty a twenty-round magazine in less than two seconds. They are more powerful and more accurate even than the AK-47 Kalashnikov, which is also widely used by poaching gangs. Rocket-propelled grenades have also been recovered, though these are not used against elephants – that kind of firepower would destroy the ivory. Rather, they are reserved for use against rangers. Law enforcement officers have also had to face American-made M-16 rifles supplied originally to the Somali defence ministry.

Against all this, rangers in some places still carry the ancient French MAS-36 bolt-action carbine, which comes complete with bayonet. 'With which,' an Interpol official told me drily, 'the entire French army lost every battle of 1940.' The defenders don't have body armour. Often they do not have tents, or ponchos or sleeping nets to keep off the mosquitoes at night. Even water bottles are in short supply so men have to drink river-water and take their chance with dysentery as well as malaria. The enemy are heavily armed units schooled in military tactics and paid to kill. In a typical twenty-year career, a park ranger has a higher than one-in-twenty chance of death in action. By 2010, a monument at the Nairobi headquarters of the KWS, which had been established in 1990, already had forty-two names on it. Another ranger recently had been murdered, and two wounded, in an ambush. In the Democratic Republic of Congo, 100 rangers are killed annually. Similar stories come from Senegal, Ghana, Zambia, Tanzania, Chad . . .

Like many African countries, Chad held an official stockpile of 'legitimate' ivory accumulated through seizures, natural deaths and the culling of rogue elephants. This was stored in the Zakouma National Park, where poachers kill many hundreds of elephants every year. In 2007 it was attacked by the infamous Sudanese Janjaweed militias. They were repelled, but three rangers died and four more would be killed later in the year. To prevent a repeat, the Chadian government burned the stockpile. But this, of course, brought little benefit to the surviving elephants. In 2006 Zakouma had a healthy population of 3,880. By 2010 it was 617 and falling.

Some other countries, including Tanzania and Zambia, took a rather different view. Rather than destroy their stored ivory, they wanted to sell it. Somehow they managed to convince European governments that legalised trade could actually save the elephant.

Well-meaning but naive, the Europeans misunderstood what they were up against. At the most conservative estimate, the amount of ivory being illegally traded each year is 100 tonnes, or approximately 12,500 elephants' worth, but that is a calculation based on seizures. The Interpol veteran I spoke to – a man with twenty-five years' experience of wildlife crime – reckons the likelier toll is the 38,000 I have already mentioned. In a market as greedy as this, a limited amount of 'legitimate' ivory was hardly likely to permanently depress the price to a level that would make poaching unviable. Indeed, it might have the opposite effect. And that is just the economic case, quite apart from the stick-in-the-craw notion that body parts from vulnerable species are legitimate items for trade. To believe that ivory, from whatever source, should be exposed to market forces requires something more than pragmatism. It is the province of the cynic.

The problem with big numbers is that they come at us every day in a plethora of contexts and their reality is literally unimaginable. They slide across our consciousness and barely register, like miles to Mars or the number of fleas on a hedgehog. But we should pause. On the day before I wrote this paragraph, two well-supported teams in the English Premier League, Tottenham Hotspur and Newcastle United, played a match at Tottenham's White Hart Lane stadium in north London. The crowd numbered 36,176 – not quite 38,000, but near enough. Imagine this number of people dead, and the size of the heap they would make. Now imagine they are not eleven-stone humans but elephants weighing up to six tonnes each. For my own benefit, I make a poor sketch of an elephant and add the dimensions of the converted cart-lodge in which I work. If a big bull stood up, it could wear the building like a shell. I do my best to visualise 38,000 of them decomposing with their tusks hacked out, but it's beyond me. And what is it all for?

The principal use for ivory in China and Japan is for *hankos* – carved cylinders engraved at one end with the owner's seal, used for stamping documents. Like all the other artefacts – chopsticks, calligraphy accessories, cups, bowls, jewellery boxes – they could just as well be made from any one of dozens of other materials, none of which would involve shooting elephants. The killings that supply the raw material are not random chance events. From trigger-finger to cash register, the process of production and supply is carefully organised. A forensic technique developed at Washington University's Center for Conservation Biology exploits the fact that elephants in different populations vary slightly in their DNA. By analysing dung samples, scientists have been able to draw a 'DNA map' of Africa, against which samples of recovered ivory can be matched. The results are unequivocal. Attacks are carefully targeted and made to order. This research has enabled Interpol to build a much clearer picture of the way the business is run. The kingpins are dealers in the Far East, who place their orders with middle-men in Africa, who control the gangs, many of which are supplied by militias, Somali warlords or rebel armies. It is probable that the poaching itself funds the militias and thus ensures the continuance of violent tribal and political strife.

From the killing zones the ivory moves quickly into intricate distribution networks. It will not be shipped from the country of origin. One consignment from Zambia was found to have travelled via Malawi and Mozambique to be exported from South Africa. My Interpol contact spoke of a 'shell-game shuffle' of multiple exit ports and indirect routing via Vietnam, the Philippines and Thailand. So complex is the system that consignments sometimes pass through the same port twice. In the rare cases of seizure, the smugglers want their supply bases to be untraceable. But the risks are small. For police forces preoccupied with murder, robbery and drug-running, wildlife crime is not a high priority. At the ports even honest officials pay scant heed to what is crossing the dock. They are intent on intercepting guns and drugs coming *into* the country. What goes out is not their problem.

At the receiving end a port official may be encouraged to take an early lunch-break, but it hardly matters. Ports in the Far East daily move tens of thousands of containers, and less than 2 per cent of them are inspected. The ivory is waved through, usually hidden beneath some innocent cargo such as timber, soapstone or sisal, to end its journey at a factory where craftsmen with power saws, lathes and polishing machines turn it into merchandise. All of this, says the Interpol man, bears the fingerprints of well-run syndicates. 'The factory needs a management team. There has to be control of inventory, a production department, a marketing department, delivery vehicles and a sophisticated finance department capable of providing payment for illegal workers and laundering millions of dollars in criminal profit.'

Every so often the investigators get lucky. They had a notable run in the summer of 2006 when they recovered 1,094 tusks at Kaohsiung in Taiwan, 390 tusks and 121 pieces of cut ivory at Hong Kong, and 608 pieces of raw ivory, equivalent to 260 tusks, at Osaka. The raw tusks alone represented 872 elephants. DNA showed that all were from East Africa, and that the Taiwanese haul was from Tanzania. To say that the criminal business was booming is to understate the case by an order of magnitude. CITES' official monitoring service, the Elephant Trade Information System (ETIS), reported in 2009 that trafficking had doubled in a year. Hauls included 6.2 tonnes from Hai Phong, Vietnam, in March 2009; 3.5 tonnes from the Philippines in May 2010; 2 tonnes at Bangkok international airport in August the same year, and the 1,768 kilos recovered in Operation Costa. In November, poachers extirpated the entire elephant herd in Sierra Leone's Outamba-Kilimi National Park.

I am writing this on 14 February 2012. To check the trend, I look up the figures for January. Early in the month two Chinese men in South Africa were caught with 'several elephant tusks and ivory goods'. On the 6th, customs in Port Klang, Malaysia, seized 494 kilograms of raw tusks, bubble-wrapped and hidden among used tyres and flooring materials in a container shipped from Cape Town. On the 14th, the UK Border Agency found ten carved ivory ornaments and a hippo's foot in the luggage of a woman arriving from Zambia. On the 26th in Polokwane, South Africa, police arrested a man in possession of four tusks, three rhino horns and firearms. There were reports of wide-spread poaching in Zimbabwe, where thirty elephants had been found dead in Mana Pools National Park. Worldwide, more than a hundred elephants were being killed every day.

All businesses involve an element of risk, usually based on calculations of supply and demand. For criminal enterprises that can switch commodities – drugs, firearms, ivory, people – with a click of the fingers, the orthodoxies of market economics are not an issue. Profit is balanced against the risk of jail, not the risk of bankruptcy. And this is what makes ivory so attractive. Smuggling tusks instead of drugs earns similar profits for a fraction of the risk. In April 2000, a Japanese government official was caught smuggling 492.3 kilograms of ivory – at least fifty-five elephants' worth – into Osaka. He was fined 300,000 yen, equivalent at the time to 2,700 dollars, or less than 2 per cent of the value of the ivory. By contrast, in the same city two years later a British man was jailed for fourteen years for smuggling 10 pounds of ecstasy and cocaine.

It was late in 2008 that CITES took what it insisted was a rational decision, rooted in good intentions and grounded in logic. If it couldn't beat the criminals, it reasoned, then it would join them. It would set up in competition and, by so doing, bring the ivory market down. Four southern African countries – Botswana, Zimbabwe, Namibia and South Africa – would be allowed to auction their stockpiles. By flooding the market with 108 tonnes of 'legal' ivory, they would put the poachers out of business. The decision was met with disbelief. Richard Leakey, the distinguished conservationist and former director of the Kenya Wildlife Service, warned CITES that selling to the highest bidders would drive the price up, not down, and poaching would become even more profitable.

Disbelief turned to consternation when CITES chose as its trading partners the very countries, China and Japan, that sustained the illegal markets. Leakey pointed out that Chinese traffickers had been convicted of smuggling ivory from twenty-two of the thirty-seven African countries that still had elephants. The inevitable result of Chinese involvement, he predicted, would be to open up the illegal markets. In England Will Travers, chief executive of the Born Free Foundation, wrote to the then environment minister Joan Ruddock, urging her to oppose the sale. She refused. 'It is your opinion that this sale will fuel illegal poaching,' she wrote back, 'it is ours that it will not.' Selling to China didn't bother her either. 'The EU delegation was satisfied that China had met the criteria which meant establishing robust controls to ensure that only legal ivory is imported . . .'

For a long time afterwards the British government's Department for Environment and Rural Affairs (Defra) refused to accept that the sale had backfired. 'The evidence which would show whether or not that decision was the right one isn't available yet,' it said in December 2009, and referred me back to the Elephant Trade Information System. In fact, I already had a statement from ETIS which it had published a month earlier.

While it agreed that the effects of the sale were not 'clear cut', it acknowledged that a 'remarkable surge' in ivory seizures had suggested 'increased involvement of organised crime syndicates'. It also found that illegal ivory typically 'follow(s) a path to destinations where law enforcement is weak and markets function with little regulatory impediment'. In case anyone doubted who it had in mind, it added: 'China . . . faces a persistent illegal trade challenge from Chinese nationals now based in Africa. Ongoing evidence highlights widespread involvement of overseas Chinese in the illicit procurement of ivory, a problem that needs to be addressed through an aggressive outreach and awareness initiative directed at Chinese communities living abroad.'

In the real world of blood and bullets, there is little patience with the equivocators. In late 2009, Patrick Omondi, head of conservation at the Kenya Wildlife Service, checked off the figures for me. 'In 2007 Kenya lost forty-seven elephants. In 2008 it was 145. This year [2009] it is 220, the worst since 1989.' Shortly before Operation Costa, more than half a tonne of ivory was intercepted at Nairobi's Jomo Kenyatta airport. Destination: China.

'The legal sales have led to illegal killing across the continent,' he said. 'We are doing all we can, but we have seen an increase in demand so high that it puts a lot of pressure on our law enforcement. The logic behind the sale – that it would satisfy demand in China and Japan – was not true.' Nothing much has changed. In July 2011, Kenya's president, Mwai Kibaki, ceremoniously set fire to five tonnes of contraband ivory, 600 elephants' worth, with a black market value of sixteen million US dollars, hoping to convince the syndicates that they couldn't win, and to encourage other countries to follow his example. Hope and experience, however, are on different trajectories. Only a few days earlier, poachers in the north of the country had been caught with forty-one tusks. In January 2012, another Kenyan game ranger was shot and killed. And so it goes on.

The one good sign is the return of something nearer sanity in the European official mind. The international community at last has ceased to believe that it can stop the ivory trade by making it legal. For CITES, the problem has always been more difficult than it sounds. Its job as a trade organisation is not to ban sales of endangered species but only to ensure that the trade is sustainable – an economic proposition, not an ethical one. What this means in effect is that trade goes on until someone can prove that it shouldn't. The issue of ivory finally came to a head at a meeting in Qatar in March 2010, when two important 'range states', Zambia and Tanzania, argued that their elephants should be downgraded from CITES Appendix I, which bans all international trade, to Appendix II, which allows it 'subject to strict regulation'.

This is classic doublespeak. Law enforcement in Africa is stretched beyond breaking point. In its wide-open spaces the keenest eyes belong to vultures and strictness is a concept observed only by the tracking of the sun. By 2010, poaching in Zambia and Tanzania was out of control. According to a park official quoted by the Tanzanian newspaper *ThisDay*, the country's Selous Game Reserve was losing fifty elephants a month. The same official made an accusation which, if true, would be a benchmark in the history of cynicism. 'Sometimes,' he said, 'authorities torch the carcasses of elephants that have been killed by poachers to conceal the truth about the extent of the problem.' There was no way of confirming this, but Zambian and Tanzanian ambitions were nothing if not transparent. They wanted to sell their stockpiles – more than 111,000 kilos of ivory – presumably into the very same channels that were used by the poachers. All the same old arguments were rolled out. It would flatten the market. The money would be invested in conservation. The good guys would win, the bad guys would lose, and hippos would dance the polka.

The Kenyans, who had suffered a 400 per cent increase in poaching since 2007, were outraged by their neighbours' attempt to exploit a 'malicious loophole'. The whole idea of national 'ownership' was false. Elephants flow as easily across boundaries as air and water do. Patrick Omondi told me about seven elephants killed near the Kenya–Tanzania border. 'Five were on the Tanzanian side, and two on ours,' he said. 'It is difficult to say, this is a Kenyan elephant, or this is a Tanzanian elephant.' The governments of the DRC, Mali, Rwanda, Ghana, Liberia and Sierra Leone, all members of the twenty-three-strong

One of the lucky few. This young elephant, photographed at the Ol Pejeta conservancy in Kenya, has better protection than most

African Elephant Coalition, joined Kenya in opposing the renewal of trade. With a horribly apt sense of timing, within a few weeks of Tanzania and Zambia submitting their applications, the government of Sierra Leone announced that its last few elephants had all been killed.

And where stood the UK? How would it use its influence at CITES? I asked the question, and back came the answer. The European Union as a whole, said Defra, had yet to agree a position. It was waiting for 'scientific evidence'. By whom would that evidence be supplied? Who else but those best placed to provide it. The national governments of Zambia and Tanzania.

For the elephant, for Africa, it was a season of crisis. The cost of a liberalised ivory trade would have been far worse than just the demise of the world's biggest land animal. Tourism is the lodestone of the Kenyan economy. Its imports must be paid for in hard currency, not in Kenyan shillings, and it is tourism that turns the vital dollars and pounds. The tourist trade collapsed during the previous poaching surge in the 1970s and '80s, and it would have collapsed again if animals disappeared or if the parks became too dangerous.

The animals contribute in other ways too. When people in Nairobi turn on their taps, it is the elephants they must thank for their water. Pools, reservoirs and aquifers are fed by water trickling from forests that absorb and release moisture like sponges. Healthy forests need diversity, and diversity needs sunlight. It is the elephants pushing down trees that open the canopy and prevent the development of monocultures vulnerable to pests, diseases and fires, and much less efficient as sponges. Some plants won't even germinate unless their seeds have passed through an elephant's gut.

The tangled complexities of politics, economics and ecology reflected the eternal conflicts. Today versus tomorrow, greed

versus conservation, death versus life. There was, too, a single
irreducible truth. One way or another, elephant poaching would
cease. Only CITES and its member states could determine
whether this would be because they had beaten crime, or because
there were no animals left to shoot. It's a question that hangs
over Africa like a witch doctor's curse. The enemy is formidable,
the future unknowable, but at least the enemy no longer includes
Britain and its European partners. After months of prevarica-
tion, the then environment secretary Hilary Benn announced
that the UK would, after all, oppose the ivory trade. Sense was
seen; justice done. At the CITES conference in Doha, the
would-be ivory traders of Zambia and Tanzania were overwhelmingly outvoted.

For the poachers, however, it was a political abstraction that
left them untouched. Whatever the law said, it was business as
usual. Within hours of the debate at Doha, twenty-four uncut
tusks were intercepted by the Spanish Guardia Civil near
Barcelona. Four weeks later, Thai authorities seized 296 tusks
at Bangkok international airport. With the hideous irony we
have come to expect, they had been shipped from the host nation
of the CITES conference, Qatar itself. In May, Interpol mounted
another series of raids across Botswana, Namibia, South Africa,
Swaziland, Zambia and Zimbabwe, seizing 400 kilograms of
ivory and rhino horn, closing down a factory and making forty-
one arrests. In the same month, still only a few weeks after
Doha, forty-eight tusks were intercepted on a main road near
Nairobi, and a tonne of ivory was uncovered in a load of African
snails at Hai Phong in Vietnam. In October, Cameroonian
authorities raided two hunting camps and found half a tonne
of elephant meat, twelve tusks and assorted weaponry. In
November, 384 tusks were seized at Hong Kong, having traced
a circuitous route via Malaysia from Tanzania. In December,

more than 100 kilograms of ivory was seized and sixteen dealers arrested in Gabon, and two Singaporeans caught with 92 kilograms at Jomo Kenyatta airport. And so it has proceeded ever since. Each arrest, each seizure, is a small victory for the enforcement agencies. But each scratch of the surface confirms the lethal vigour of an exponentially bigger trade that operates with little interference from the law. For every tusk intercepted, ten escape the net.

I'm afraid that my voice in this chapter may have become somewhat shrill, but it is impossible not to be aghast. Impossible not to reflect on the bloodbath, so proudly chronicled by Roualeyn Gordon-Cumming, that set the tone for all that was to come. I am anticipating with pleasure my visit to Ol Pejeta, but even as I pack my bag I have a hideous image in my mind. A winning portfolio in the 2012 World Press Photo awards, by the South African photographer Brent Stirton, includes a picture of a female rhinoceros, nose to nose with her mate. She has no horn. Somehow, four months before the picture was taken, she survived an assault by poachers in which the horn and a section of bone were removed by chainsaw. I don't know how they did it, and I don't think I want to.

CHAPTER EIGHT

Ol Pejeta

Quite unexpectedly a long-forgotten question is brought to mind. In my early childhood I lived in a village in south Bedfordshire, surrounded by sprout fields and piggeries which gave the air a permanent faecal tang. To this generally agreeable, porky aroma would be added once a week the stomach-churning waft of the 'lavender lorry' on its round of the cesspits. Sometimes this malodorous vehicle would be so overfilled that it would spill over and mark its passage with a glistening trail of brown along the road. It was not this, however, that pricked my curiosity. The only mystery about sewage was why people called it 'night soil'. It was, rather, the dogs, which back then in the 1950s used to speckle the pavements with faeces of pure white. Being accustomed to this, we boys saw nothing odd in it and called it 'dog chalk'. Until now this distant phenomenon had slipped from my memory, and so I had failed to wonder why modern dogs do things so differently. But now I think I may have an answer.

To find it I have had to make a journey of some 4,500 miles from my home in eastern England to a parched high plateau beneath the misted peaks of Mount Kenya. In the dog-chalk days this was a world I knew only from the limpid grey images of Armand and Michaela Denis, glimpsed through my parents' veneered Ekco television. Then it seemed impossibly far away, exotic and bathed in dangerous glamour. Now it has to be filtered through more than half a century of televisual over-familiarity. It worries me. Perhaps a life in environmental journalism has left me incapable of surprise. Will I rediscover my hidden *naïf*?

My visit to Kenya has coincided with a leaf-storm of articles hailing Africa's 'new economic miracle'. A writer in Kenya Airways' in-flight magazine complains about 'knee-jerk journalism', by which he means the reflexive habit of referring to Africa as an economic basket case, and of sanctimonious hand-wringing about poverty, violence and corruption. There may indeed be a new spirit of entrepreneurism, and it may be true that more African-run companies are earning profits and enriching a new middle class. In the leafy Nairobi suburb of Karen you could easily believe it. Here, spread across the former coffee farm once owned by Karen Blixen of *Out of Africa* fame, vast mansions stand in gated five-acre plots. Wealth drips like honeydew from the trees. As you would expect in this sub-tropical simulacrum of Surrey, there is a Country Club with ornamental lakes and a championship-standard golf course.

But to reach it you have to brave the linear snakepit of the Mombasa road, a just-about-moving tailback of ancient Japanese saloons trailing petticoats of rust and hassled by the even more rackety private minibuses – the notorious *matatus* – which are as close as Nairobi gets to public transport, and whose unsignalled lane-switching could hardly be more alarming if the drivers wore blindfolds. On your right as you crawl out of the city, a vast rust-coloured stain spreads as far as you can see – the Kibera slum, whose tin-roofed shanties according to wildly varying estimates are home to between 235,000 and one million people, crammed in at a density of perhaps 200,000 per hectare, living in conditions that would struggle to be called medieval.

A long-term slum-clearance scheme is under way, but the average man on the *matatu* would boggle at any thought of an economic miracle.

Out in the country things are not much different. From Nairobi's local airport, Wilson (named after Mrs Florence Kerr Wilson, a feisty widow who set up Kenya's very first airline in 1928 with a two-seater Gypsy Moth), I am bounced around various dusty, miles-from-anywhere airstrips before touching down eventually at Nanyuki, in Laikipia county. Here I am met by Andrew Odhiambo, from Kicheche Camp, an hour away at the Ol Pejeta Conservancy, who will be my guide for the week. Being exactly on the equator, we are constantly dodging back and forth between hemispheres, defining Latitude Zero with a rooster-tail of grit. This is such hard country that you could say it's an economic miracle that anyone scratches a living from it at all. There are many flimsy churches and evidence of prayers in urgent need of answer. Shopping centres are dirt roads edged with tumbledown *dukas*, more like sheds than shops, which are approached through a moonscape of axle-deep potholes. Arable fields bear scrawny remnants of wheat, or sun-scorched grass forlornly scoured by a few ribby cows, sheep and goats (this is the end of the dry season, when the entire country gasps for rain). A community of Maasai pastoralists occupies a row of huts that would not look out of place in Kibera. Miracle of tenacity and last-ditch human resourcefulness? Yes, probably. Economic miracle? I don't think so. Not here. Not yet.

Living in England, I am no stranger to wide discrepancies between rich and poor. But poverty in Britain is relative, not absolute. There is a loss of dignity, self-esteem, opportunity and enjoyment, for which liberal opinion can feel ashamed, but there is not usually a threat to survival. The contrast with Africa is stark. The incongruity of so much of the world's anxiety, including my own, being focused on the well-being of *animals* is a searching test of moral perspective. Right now I am struggling to cope – just as Alfred Russel Wallace and Julian Huxley must have done – with a blitzkrieg of the senses. As I said, I am no kind of Africa hand. I am not even a very competent traveller, being unable to shake off anxieties about missed connections, lost baggage, misread timetables and sniffer dogs (on a previous trip, one of these nailed me at Johannesburg for carrying piri-piri in my luggage). In the tropics I fret about malaria pills, sunblock, insect repellent and how I'm going to get home again. At Ol Pejeta, all that evaporates like spit on a barbecue. At the penultimate airstrip in my sequence of low aerial hops, a herd of elephants was stripping the trees next to the runway. As the horizon breaks open at Ol Pejeta, the first thing I see is a giraffe, lolling through the acacias with that strange Anglepoise gait that it shares with camels. And suddenly, right here, the opposing worlds of the mundane and the imagined merge into a single moving frame. I am struck daft, as if by a bolt from boyhood. A real giraffe in real Africa! Like hundreds, thousands, millions before me, I am overwhelmed. My scepticism withers and is forgotten; the *naïf* steps blinking into the sunlight.

And dog chalk? Andrew is a brilliant naturalist who will not leave the smallest detail unexplained. Even as he drives, his eyes are scanning the horizon, the tree-line, the very dust in the road where the imprints of hoof and paw tell the story of the day. One morning we stop by the bleached skull of a buffalo ill-met by lions. Not far away is what looks like a pile of dog chalk. 'Hyena,' he says, crumbling it with his shoe. The hyena is a remarkable animal which deserves better than its pejorative reputation as scavenger and thief. It is a skilful hunter in its own right. What it lacks in speed it makes up for in stamina, running for miles in pursuit of prey that might at first outpace it. A zebra, for example, will be run to exhaustion until it stands defenceless and surrenders to its fate. It has to be said that the fate is not a good one. Unlike lions, which kill usually by suffocation, hyenas begin their meal while the dish is still on its feet. The power of their jaws is terrifying. If there were any nutritional value in granite, then they would sink their teeth into it. As it is, every part of the victim is ground up and swallowed. The faeces are white, Andrew explains, because of the powdered bone in them, and it is now that I am reminded of my Bedfordshire childhood. Back then, long before supermarket aisles were lined with tinned gravy dinners, dogs were given bones to chew. The chalky pavements are explained.

Bedfordshire, however, does not stay long in the mind when you've got wild Africa in your face. It seems almost absurd to have so much dished up at once – so over the top that it makes me laugh, like a child at a fairground. How can this be *real?* There are fences around the conservancy, but the area within them is huge, 350 square kilometres, and they are there to channel the movement of wildlife, not to obstruct it. Well-used corridors through the fencing allow animals to move freely in and out, while steering them away from villages and farms. This crucially prevents the 'island effect', a weakness of nature reserves isolated from their surroundings which leads to the local extinction of some species and over-population of others. As its chief executive, Richard Vigne, will explain, Ol Pejeta *accommodates* wildlife. It doesn't farm it. What I am seeing therefore is recognisably the same place that Alfred Russel Wallace saw in the 1870s. In fact, I am very likely seeing even more then Russel Wallace did. Unlike him, I have an expert guide at the wheel of a Toyota Land Cruiser, an indefatigable, go-anywhere hyena on wheels in which Andrew can deliver me into close proximity with almost any animal of my choosing. But that 'almost', I now confess, means the sad exclusion of golden moles. Truthfully, deep down, I have always known I wouldn't see one and so have delayed the question until after my arrival. In that rather cowardly way I have kept alive a slender thread of hope and justified a thirty-hour journey to the middle of Africa. But I block the thought. The mole, I guiltily acknowledge, was always an excuse – a detail, seductive but arbitrary, that would draw me into the bigger picture. And now here it is, a picture so enormous that I can't take it all in. *Calcochloris tytonis* and its tribe can wait a bit longer. In any case, I feel it is here in spirit, represented in its absence by myriad tiny scurriers and burrowers that only civets, hawks and owls can see. At breakfast one morning a tiny striped mouse boldly darts out to feed on a handful of muesli – the smallest animal I see, and my mole by proxy. The economic principle of Fritz Schumacher's *Small is Beautiful* may now be derided by the more macho kind of free-marketeer, but it holds good in nature. Small is not only beautiful but, as my pursuit of the mole will make abundantly clear, it is also essential.

Andrew is a great talker. He explains the byzantine intricacies of African politics and tribalism – he is of the Luo tribe, though the people of the district are mostly Maasai and Kikuyu – and he has an insatiable appetite for news. Who did I think would win the Republican primaries currently being fought in the USA? Mitt Romney, I say, and he agrees, but he wants to know what a Republican victory would mean for American minorities. I wish I could tell him. Andrew's passion for wildlife matches even that of my old friend (and genuine Africa hand) Brian Jackman, on whose advice I am carrying a new pair of 10x42 binoculars and a fleece to keep off the evening chill. Being on a plateau I get no sense of elevation, but we are actually 2,000

metres above sea level, slung between the northern slopes of
Mount Kenya and the Aberdares. I do at least manage to astonish
my guide by showing him what Europe looks like at this altitude.
The photographs on my camera were taken from a ski station
high in the Swiss Alps at Verbier, in mid summer but still
hemmed in by snow-capped peaks.

I learn soon to abandon my own swivel-eyed scouring of the
plains and rely on Andrew's seemingly supernatural ability to
read nuances of light and shade. He has sharper senses than
any other human I have ever met, including even Jackman and
the angling writer Brian Clarke, who once pointed at a ripple
on the River Test and predicted to within a few ounces the
weight of the trout that was causing it. Andrew has wraparound
eyes and ears. Distant pinpricks, invisible even to my 10x42s,
turn into rare Jackson's hartebeests. Faraway murmurs swell
into waterbuck. Time and again, quietly and carefully, he plants
me within a how-do-you-do of elephant, buffalo, rhino, giraffe,
eland and their supporting casts of jackals and hyenas. On one
memorable afternoon he ushers me into the presence of a hippopotamus. Cheetah and leopard require more luck than comes
our way (though both are here in numbers); otherwise nothing
escapes him. A jackal trots past with something in its jaws – the
head of a baby hyena, Andrew says, 'very unusual'. A dot of sky
blue becomes the scrotum of a vervet monkey (the colour is
what separates the men from the boys – adolescents display an
immature shade of green). Young baboons lark and dart through
a fever tree, playing a kind of Kenyan roulette with gravity while
their elders hunch in the branches like giant rooks.

One morning we set out at six fifteen to look for lions. For the
first couple of hours we have no luck – or, rather, our luck is of
a different kind. Serendipitously in the dawn light we find a black
rhinoceros and her calf standing rock-still in the bush. There are

giraffes wallpapered against the lightening sky, buffaloes trudging head-down across the plain, a group of oryx. We meet elephants, warthogs, a hyena carrying the leg of a gazelle, eagles, ostriches and uncountable zebra. But no lions, and we are getting hungry. On a curve of the Uaso Nyiro River, away from the trees, Andrew sets up the table for breakfast. Red checked cloth, cereals, yogurt, sausages and bacon, vegetable pies, pickles and preserves, fruit, tea, coffee . . . This time we have no need of Andrew's enhanced sensory perceptions. The sudden deep, guttural cough, *Wugh!*, is no distant murmur. It is nearby, urgent and loud. Twenty metres away a lioness pads out of the trees, glances at us without interest and lopes off along the river.

We follow in the Land Cruiser for as far as we can, before she melts into the bush still calling for her lost companions.

Uninvited guest – the lioness brought a sudden end to breakfast

Encounters in Africa, I gather, are often like this. Plains game you can guarantee, any time of day or night. Zebra, Thomson's and Grant's gazelles, warthog, giraffe – you can't avoid them. Lions, however, are different. You can improve your chances by knowing where to look, but you still need a bit of luck. A couple of days later Andrew stacks the odds in our favour. Using a radio tracking device, he locks on to a collared lioness lying somewhere deep in the bush. Even so, in thick undergrowth she is not easy to find, and I congratulate myself – the only time it happens – for spotting her before Andrew does. She is hidden in deep shade beneath a tree, together with another lioness and two well-grown cubs. Like the breakfast lioness they only cursorily note our arrival. Four heads pop up at the sound of the engine, then flop back down again to doze. Their bellies are full; their eyelids heavy, hunger forgotten. Barely a hundred metres away, an impala skips across a clearing, upwind and unaware of its luck.

Another day, we rumble across the plain to the conservancy's airstrip. It is heavily grazed and, at ground level, difficult to distinguish from the land around it. A small Cessna has touched down and is parked among a group of gazelles. Down from the cockpit steps Richard Lamprey, Fauna & Flora International's technical specialist for Kenya, Uganda and Tanzania, who has flown up from Nairobi to meet me. FFI's fingerprints are all over Ol Pejeta. The land was once a 40,000-hectare ranch owned by the mining and farming conglomerate Lonrho, which was put up for sale in 2004. Conventional cattle ranching by then had become difficult. Controlling wild animals had been made all but impossible by a hunting ban, and by an influx of elephants from the drier country to the north which had destroyed most of the fencing. Decreasing productivity and rising costs were driving ranchers into insolvency.

For wildlife, and for the conservationists who cared for it, it was a situation that presented both danger and opportunity. The danger was that the land would be sliced up into plots. As movement through the area was vital to the flow of animals across the plateau, this would have critically reduced its value for wildlife. The opportunity was to buy and save the land for wildlife, and at the same time to grow the local economy. This is the twin-track approach upon which the future of endangered species ultimately must depend. The late Christopher Hitchens pointed out an oddity of the English psyche – evidenced by involvements in places like Greece and Spain – which leads the queen's subjects to show more enthusiasm for other people's patriotism than they do for their own. For conservationists in particular, this has tended to extend not just to other peoples but to other species, whose 'rights' are promoted over man's. But there is a fatal flaw in this. Animals are killed or displaced for a reason. Their persecutors expect to profit by it, and conservation is not popular where communities feel their interests are secondary to those of the wildlife. The substitution of the word 'poaching' for 'hunting' in the language of the law is a perfect example of conflicted priorities. I am reminded of the two prisoners in Mozambique's Gorongosa National Park who were punished for shooting a warthog. It is a simple fact. Conservation cannot succeed without popular support, and people as well as animals need to see the benefit.

As Richard Lamprey explains, FFI is not in the business of owning land. It *is* in the business of encouraging purchases where land is of value to wildlife. At Ol Pejeta it managed to secure funding from the Arcus Foundation, a private charity set up by an American philanthropist, John Lloyd Stryker. Stryker had a particular passion for great apes – importantly for him, Ol Pejeta already had a refuge for chimpanzees – but he shared

FFI's broader vision and wanted to help. So it was that by late 2004 Ol Pejeta belonged to FFI, though its ownership would be short-lived. By the end of 2005 the land had been transferred to a locally based non-profit company, the Ol Pejeta Conservancy, staffed and run largely by Kenyans, though FFI would remain an active partner. Richard Lamprey is a frequent and welcome visitor.

At the conservancy's somewhat ramshackle headquarters, known as 'Control', he introduces me to the research officers Samuel Mutisya and Nathan Gichohi, whose presentation is a statistical tour de force – breakdowns of animal populations by species, sex and age, breeding rates, causes of death, movements in and out of the conservancy. It builds into a minutely detailed portrait of a landscape exploding with life, and helps me remember why I am there. Forgetting golden moles, and setting aside the excitement of Andrew's game drives, what brought me here was the rhinoceros. No animal is more charismatic, none has a higher bounty on its head or is in more desperate need of protection. Ol Pejeta in the last year or so has lost three of them to poachers, including a rare southern white found with seventeen bullets in its body. Rangers had already de-horned it as a precaution, but the poachers still hacked off its face to get the small amount that remained. There has been a human cost too. Six months before my visit, a poacher was shot dead and two others wounded in a firefight near the conservancy boundary. Now another rhino has been wounded by poachers and badly needs help. Rangers and vets will search for it next day, and Richard Lamprey and I will go with them.

The area of search is defined by the orbit of a spotter aircraft – a Piper Super Cub kept for just this kind of eventuality – which guides the team on the ground. Sensibly I'm kept away until a tranquilliser dart has done its stuff and the wounded monster has sunk belly-down with its legs folded beneath it. A green towel has been draped over its eyes and it snorts like a drunk in its sleep. Any living thing this big would be impressive, but the rhinoceros with all its primitive power is awesome, like a piece of living geology. It is not, however, immune to the high-velocity bullet. This one, by great good luck, was a misdirected shot that passed through the left foreleg, missing the bone but leaving the animal badly lamed. I wonder what Roualeyn Gordon-Cumming would have thought of such a scene. What would he have made of an age in which animals were so fiercely defended and men like him, whose preferred view of nature was across a gunsight, were vilified as criminals?

I can't deny that I enjoyed Gordon-Cumming's yarns. Had I discovered him earlier, he might even have been a boyhood hero. But his are not the eyes through which modern Africa can be viewed. The challenge for conservationists is to bring down the market, to make it literally true that a charismatic wild animal is worth more alive than dead. Standing there, surveying the fallen giant, I have a fantasy. If I could bring alive just one historical figure and share with him what I can see, it would be Albrecht Dürer, whose famous woodcut of a rhinoceros, made in 1515, is still the most beautiful 'likeness' of the animal ever made. 'Likeness' needs quotation marks because Dürer himself never saw a rhino, but made his drawing from a written description and a sketch by someone else. The image is recognisably of a one-horned Indian rhinoceros, but one that seems to be wearing something like medieval horse-armour, complete with rivets and tooled bodywork. Inaccurate it may be, but it brilliantly captures the monstrous strangeness and physical enormity of an animal that ancient bestiaries conflated with the unicorn.

Five centuries of increasing familiarity have done nothing to reduce its impact, on the imagination or on the eye. The details may be awry, and the species may be wrong, but Dürer's image distils to its essence the spirit of the collapsed behemoth that now lies before me. I am, I realise, the only one just standing and taking photographs. Everyone else is lending a hand, bracing themselves to heave the animal over so that the vets can reach the wounded leg.

Afterwards when they have cleaned the wound, administered slow-release antibiotics and the antidote to the tranquilliser, we watch from a distance as the patient lurches to its feet, remembers where it is and limps off into the acacias. Like every other rhino in the conservancy, it has a fortune on its face and will have to take its chances. At least in Gordon-Cumming's day rhino horn was valued only as an exhibit. Now it is a commodity. Most of the specimens in European museums were collected in the late nineteenth or early twentieth centuries, and for each institution one was usually enough. For the traditional medicine trades of China and south-east Asia, one is never enough. I am loath to correct Julian Huxley, but there is one thing that he seems to have got wrong in his influential pieces for the *Observer* in 1960 – a mistake since repeated by many others. It is not as an aphrodisiac that rhinoceros horn is prized. In fact, this seems to be the one benefit not to have occurred to the ancient herbalists whose word in these matters gives quackery its kernel of faith. In most horned animals, the horn has a bony core within a sheath of keratin (the same stuff from which hair, hooves and fingernails are made). Rhinos are different. There is no bony core, so the horn is solid keratin. Powdered and dissolved in water, it has magical properties sufficient to replace the entire stock of a western pharmacy (though if this is true, of course, it would make as much sense to chew your fingernails or eat your haircut).

The rhino shot by poachers at Ol Pejeta. Vets and rangers have to turn it over so they can treat the bullet wound in its leg

Victims of this historic scam swallow it to relieve themselves of fever, rheumatism, gout, typhoid, carbuncles, snakebite, headaches, nausea, hallucinations and daemonic possession. In Vietnam, where demand is rising fast, the list extends also to hangovers and cancer. Apart from a flagging libido, the only things it seems unable to cure are the credulity of its purchasers and the cupidity of those who keep them supplied. Huxley was right that rhino horn fetched more than 'the best ivory', but that was fifty years ago. Now it has exceeded even gold. On the day I checked (20 February 2012), the price of 24-carat on the London market was £35,165.70 per kilo. In China and Hong Kong, rhino horn was fetching £40,000, and there were reports of prices as high as £60,000. Look no further for the reason rhinos in Africa are an endangered species. Look no further for the reason African officialdom is so vulnerable to the outstretched palm. Who *wouldn't* be tempted by such sums?

Like all successful industries, crime is in a perpetual state of development, forever alert to new opportunities. Coincidentally, at lunchtime on the very same day that I made my Hong Kong price-check, four thieves walked into the Castle Museum in Norwich, just 30 miles from my home, and tried to snatch the stuffed head of a black rhinoceros, complete with horns, that had been in a glass case there since 1911. In fact, a whole new crime wave was breaking out not just in Britain but right across Europe. In June 2011 the EU's criminal intelligence agency, Europol, warned that an Irish crime group was diversifying from its already extensive criminal portfolio into the theft of rhino horns from museums. The gang-members were infamous hard nuts involved in fraud, robbery, money-laundering and drug-trafficking in North and South America, South Africa,

China and Australia as well as Europe. And they'd had a brainwave.

Horn from African wild rhinos was one of the most valuable commodities on the international market – so valuable that men were prepared to risk their lives to get it. But why face all the dangers of a shooting war when museums throughout Europe were packed with stuffed specimens, including horns, in fragile and usually unguarded cabinets? In criminal terms it was a no-brainer. A fortune was there for the taking.

And take it they did. From Sweden in the north to Spain in the south, Portugal in the west to Hungary in the east, natural history museums were targeted like banks in the Wild West. By the time Europol issued its warning, there had already been two raids in the UK. In February of that year a stuffed rhino head had been burgled from an auctioneers at Stansted Mountfitchet in Essex. It was later found, minus its horn, in a ditch. Next to be hit, in May, was the Haslemere Educational Museum in Surrey, where thieves broke in at two o'clock on a Friday morning. The museum holds 240,000 specimens, one of the largest natural history collections in Britain, but only a single item was taken – the mounted head of a black rhino, with both horns intact, which had been brought to England from Kenya (then British East Africa) in 1913 and had been on display since 1929. A few days after the Europol warning, the thieves turned their attention to Ipswich in Suffolk. Their target was Rosie, a one-horned Indian rhino that last drew breath probably some time in the late nineteenth or early twentieth century. She might have been shot in the wild, or died in a circus or zoo. No one knows. What's certain is that in 1907 the Natural History Museum in London sent her to the Ipswich Museum in return for a stuffed pig and £16. She made the journey on a horse-drawn wagon from which it took ten men two hours to unload her. Over the years she became a local celebrity. The name Rosie was conferred after a competition in a newspaper, and a drawing of her by the artist Maggie Hambling became the museum's definitive icon. For 104 years Rosie stood there, admired but unmolested. Then, in the middle of a July night, thieves broke in through the back door, wrenched off her horn, snatched an African two-horned black rhino's skull from the top of a showcase and made off with them. Bearing all the hallmarks of well-rehearsed professionals, they were inside the building for just three minutes.

As museums across the continent began to wise up and strengthen their nocturnal security, so the criminals abandoned the burglar's stealth in favour of smash and grab. With the alarms switched off and replaced by unarmed attendants not famous for their muscularity, galleries looked a softer touch in the daytime than they did at night. At Drusillas Animal Park, near Alfriston in Sussex, between four fifteen and four thirty on a late-summer afternoon, raiders broke into a display cabinet and made off with the horn of a black rhino on loan from Brighton's Booth Museum of Natural History. Ironically, the horn was part of an educational display about CITES (the Convention on International Trade in Endangered Species), which outlaws international trade in rare animals. The items in the cabinet were all things that people were urged not to buy, including ivory, coral, seashells, furs, turtle-shell, snake and crocodile skins as well as rhino horn. To avoid its becoming a criminal rather than educational resource, the display had to be removed.

Everywhere the thieves grew bolder and more contemptuous of security. In a lunchtime attack on the Museum of Hunting and Nature in the Marais district of Paris, two men used what police described as a 'paralysing gas' to subdue guards and help them get away with the horn of a rare South African white rhinoceros. Tear gas was also used in a raid at the zoological museum in Liège, Belgium. On a Saturday morning at the Museum of Natural History at Gothenburg in Sweden, men armed apparently with an electric saw smashed a glass case and lopped off the horn of a stuffed rhino. In March 2012 a British man was held on suspicion by German police after what they described as an 'unbelievably audacious' raid in Offenburg, during which two people distracted museum staff while others climbed on a display case, took down a rhino head from the wall and smashed off its horn with a sledgehammer. By the end of that month there had been fifty-eight thefts from fifteen European countries, involving seventy-two separate horns, eight entire heads with sixteen horns between them, eleven replicas and three carved rhino-horn 'libation cups'. Germany was the most popular target, with thirteen thefts and one failed attempt; France second, with eleven thefts and four attempts, and England – five thefts, one attempt – third, just ahead of Austria with four and two. Victims in Italy, I learned, included the Museum of Natural History in Florence, possible resting-place of the Somali golden mole, whose trail I would soon pick up again.

Museum exhibits, of course, were a finite resource, and curators were soon removing genuine horns and replacing them with replicas made of glass fibre or resin. As thieves tended not to be expert zoologists, this did not always put them off. In a pre-dawn raid in August 2011, a gang smashed through the front door of the Natural History Museum's Hertfordshire outpost at Tring, and hammered off the horns from a stuffed Indian rhino, originally from Cooch Behar, and the head of a white rhino from the former North East Mashonaland, now Zimbabwe. Both were collected around 1900 and had been in the museum since 1939, and both had been fitted with valueless horns moulded in resin.

White rhinos at Ol Pejeta – they have been dehorned to deter poachers

With the supply from museums now depleted, concern began to switch to zoos. So far as I could discover, no rhinos had been killed in European zoos, but there was clearly a risk. In October 2011 an antiques dealer was jailed for twelve months by Manchester Crown Court for trying to smuggle two rhino horns, hidden inside a fake bronze sculpture of a bird on a log, on to a flight to China. DNA tests traced the horns back to a forty-one-year-old white rhino which had died two years earlier at Colchester Zoo in Essex. In accordance with CITES, the animal's body had been sent for incineration at an abattoir, where it was stolen and later sold into the criminal supply chain for £400. The defendant's lawyer told the court that his client was 'just a link in the chain', but there was no doubt about what he had

stood to gain. In China the horns would have been worth at least
£400,000 and probably more. This prompted the UK police
National Wildlife Crime Unit, based in West Lothian, to warn
all British zoos to be on the alert. Colchester, which still had six
adult rhinos and a calf, introduced a ring-of-steel security system
involving night patrols and an alarmed fence connected directly
to the police.

A few days ago I heard on the radio a natural scientist say that
connecting with nature made him feel 'blessed'. This is not a
word I would apply to myself – having no connection with any
extraneous spiritual entity, I have no source of blessings – but
I understood what he was driving at. The American geneticist
Dean Hamer has hypothesised a specific human gene, called
VMAT2, which conveys a propensity for spirituality. In some
individuals this would declare itself in religious faith; in others
through different forms of spiritual expression, most obviously
in the experience of music, art and poetry. To these I would
add the beauty of the physical world. At any 'beauty spot' on
a fine day you will see people silenced by the view. Our feeling
of smallness in wild places, the aura we ascribe to the place
itself, is what we mean by the 'sublime'. The word is there in
my African notebook, turned into spider-scrawl by the jolting
of the Land Cruiser but still legible in mad-looking capital
letters. Further down the page is the word 'surreal'. One longs
for the verbal palette of a Wordsworth, but under pressure of
enthralment this is the best I can manage. *Surreal* because, to
a northern European used to being thrilled by the sight of a
hare, wild Africa is the animation of a dreamscape. It is a
sublime moment when I find myself, the voyaging dreamer,
nose to horn with a black rhinoceros in its element. I am silenced
as if before some great artwork, drawn into the other's space.

I wonder what the rhinoceros knows or feels. To what extent is it self-aware? Who or what does it think I am? How does it think of others of its own species? Does it understand its own power?

I enjoy a similar moment with elephants. One glorious afternoon we find a maternal group grazing in the bush – I hear the rhythmic, toneless sound of leaves being torn even before I see what is causing it. Andrew works out the family relationships – mothers, grown-up daughters helping with their younger siblings, the tiniest of calves. One young female strolls towards us, trunk upraised but not threatening, so close that my pocket camera cannot contain her head. It is like poking a lens into prehistory.

Early in the book I mentioned the West Runton Elephant, the 600,000-year-old skeleton of a huge steppe mammoth found in cliffs not far from where I live. It roamed in Norfolk 350,000 years before the woolly mammoth appeared, yet in terms of elephant history it was an upstart. Among the Royal Society's *Biology Letters* in February 2012 was a paper bearing the kind of typically off-putting title that keeps the general reader at bay: 'Early Evidence for Complex Social Structure in Proboscidea from a Late Miocene Trackway Site in the United Arab Emirates'. But it revealed a brilliant piece of archaeological research by an international team from Germany, France, the USA and the UAE. What they had found was a fossil trackway bearing the footprints of elephants that passed along it seven million years ago. It is another of those big numbers that you have to stop and think about. *Seven million years*. The group contained at least thirteen animals of varying sizes from calf to adult, and the prints showed that they were sexually segregated, just like the herds I am seeing at Ol Pejeta. They took this route, at Mleisa in western Abu Dhabi, 6.4

million years before the West Runton Elephant lived and died, and 6.8 million years before elephants in Africa first enjoyed the company of that evolutionary johnny-come-lately, *Homo sapiens*.

So here I am, a blatant example of human intrusion, keeping close company with the so-called Big Five – buffalo, lion, leopard, elephant, rhinoceros – and comprehensively dwarfed by what scientists call *megafauna*, meaning species larger than man. It would be hard to imagine better luck. At the reasonable cost of writing a travel article for *The Sunday Times*, I am being accommodated at Ol Pejeta's very luxurious Kicheche Camp. My 'tent' is a large bungalow with veranda and all mod cons, including flushing lavatory and hot shower – tent only in the sense that the walls are made of canvas. Barely 100 metres from the balcony is an enormous waterhole – really a lake – to which come buffalo, waterbuck, giraffe, lion and elephant to wallow or drink. Meals are almost paralysingly sumptuous, and each evening while I'm enjoying my supper some kind soul pops a hot-water bottle into my very comfortable bed (as Brian Jackman warned, nights at this altitude are chilly). For safety's sake, and adding a little frisson of drama, I am not allowed to move around in the dark without a guard (and I do hear big beasts at night). The reason for this paragraph is not just to say thank you to those who put me up. It is rather to make the point that tourism in Africa has a value that stretches way beyond the privilege granted to visitors and the profits earned by tour operators. As we shall see, the tourist is a vital link in the chain of virtue that keeps animals alive and strengthens local communities.

Drifting towards slumber, I am disturbed by a riot of snarls and shrieks – a meeting of lions and hyenas, says Andrew in the morning. Another thought hinders my return to sleep. Not knowing whether to be amused or appalled, I remember that the horn stolen in Ipswich from Rosie the rhino had been preserved with arsenic. I wonder whether the patient who swallows it will live long enough to register the shock.

CHAPTER NINE

The Virtuous Circle

Disappointingly the striped mouse, *Rhabdomys pumilio*, does not return for a second breakfast. It is, apparently and surprisingly, the most common wild animal in southern Africa – a fact one must take entirely on trust. On the evidence of Ol Pejeta you could easily believe it was the zebra, of which I see uncountable numbers against just the single mouse. They are equals in beauty, though. *Rhabdomys* is about twice the size of a house-mouse and is named for the four dark stripes on its back. It is mostly vegetarian but not above eating insects. At home, my own regular breakfast is muesli, though at Kicheche I find I am not above sausage, bacon and eggs. In colonial times Kenya was British East Africa, so the appearance of a frying pan is no great surprise. Common as it may be, seeing *Rhabdomys pumilio* was pure luck. Not even Andrew could have found one to order. It cements my conviction that I shall never see a living golden mole. Even if they were here, right now, burrowing beneath my feet, how could I bring one to light? My English garden seethes with life – moles, voles, mice, shrews – which, I now reflect, I see only when they are caught by cats. I realise, too, that my knowledge of golden moles remains embarrassingly slight. People ask simple questions and I am stumped for an answer. *But you're supposed to be writing a book!* In my defence

I say that no one ever seems to have written very much about them. Later, looking for help, I will return yet again to the nineteenth century, to the portentously titled *British Cyclopaedia of Natural History: combining a scientific classification of animals, plants, & minerals. By authors eminent in their particular department. Arranged and ed. by Charles F. Partington*. This great work was published in three volumes from 1835 to 1837. On page forty of Volume Two (1836), I find this:

> This species is a very small animal, considerably less than the common mole of Europe; in consequence of its subterranean habits it is not very frequently seen; and in respect of its colour it is as perplexing as the cameleon [sic]. We believe that the real colour, that is the colour as seen in the light which is not refracted, is brown; but, different from all the other mammalia, this small animal has the same metallic reflections in its fur which are observable in the feathers of many birds, the range of these colours being from a deep golden yellow, or rather a sort of bronze red, to a bronze green; and as all animals which have the metallic reflections lose them when dead and dried, the stuffed skin of this one conveys no idea of what the living animal is like . . .

From this I deduce it is most unlikely that the 'authors eminent in their department' ever saw a living golden mole themselves. But it usefully reminds me that the object of my quest is not a live animal at all, just a minuscule hint at a species that might not even have existed. Somewhere in Florence, surely, must repose the crumbled remains of Professor Simonetta's Specimen No. MF4181, the only physical record of *Calcochloris tytonis* anywhere in the world. Even if I don't find it, I am substantially in its debt. The unseen little creature has made me think. I realise, despite the Latin binomials littering my text, that I understand little more about the classification of species than could be gained by looking up Linnaeus in an encyclopaedia. What does it *mean* to say that aardvarks are related to cetaceans, or bats to primates, or golden moles to the marsupials of Australia? How might the evolutionary process respond to an epoch so altered by man that scientists are calling it the Anthropocene (from the Greek *anthropo-*, meaning 'human', and *-cene*, meaning 'new')? For how many species will it be survivable? What *is* the value of a species? I know that a lot of very big brains have travelled the ground ahead of me, but I know also that ignorance puts me in good company. Leafing through the scientific literature, the layman is as much struck by the holes in it as by the erudition. What sparked my curiosity was not some believe-it-or-not detail of animal behaviour or adaptation. It was turning the pages of *Mammal Species of the World* and finding that an entire species was 'known only from a partially complete specimen in an owl-pellet'. *Fragment* would have been a better word.

The entry for the black rhinoceros allows no such equivocation. *Rhinoceros bicornis* was first classified in 1758 by Linnaeus himself. No doubt he would have thought it as secure in its niche as its relative the horse, though by identifying its homeland as 'Habitat in India' he was somewhat errant in his geography. *Mammal Species of the World* now rather forlornly defines its range as 'formerly' in Angola, Botswana, Burundi, Cameroon, the Central African Republic, Democratic Republic of Congo, Chad, Eritrea, Ethiopia, Kenya, Malawi, Mozambique, Namibia, Niger, Nigeria, Rwanda, Somalia, South Africa, Sudan, Swaziland, Tanzania, Uganda and Zimbabwe. Given such a litany, even empirical, unemotional science can't keep the sorrow out of its voice.

No one in Linnaeus's day would have known how many black rhinos there were, and no one would have seen any need to count them. You might as well have counted starlings. Whether by hand of God or through the twists of evolution, Nature had done an excellent job and species had settled into a harmonious if sometimes bloody state of equilibrium. The local 'carrying capacity' for a species was determined by the amount of space each animal needed, and by the balance between predator and prey. From microbe to elephant, everything was safe in its niche. Everything, that is, save for one bipedal rogue which, through God-given dominion, counted itself superior to all the others. The equilibrium of the wilderness was shattered by human intervention. One species after another was driven to a last redoubt. Many simply perished, to be forgotten like the long-tailed hopping mouse, or vainly sought by anguished resurrectionists like the thylacine. Without a determined rescue effort the black rhinoceros would have gone the same way. By 2001 only 3,000 were left in the whole of Africa. In Kenya they fell from 20,000 to fewer than 300, a rate of loss equivalent to 4.5 rhinos every day for ten years. Now Kenya is back up to 620, of which eighty-seven, roughly 15 per cent of the total, are at Ol Pejeta, the biggest single population in East Africa. As the theoretical carrying capacity here is 120, the time cannot be far away when they will spread out beyond the fence. Inspired by Ol Pejeta, twenty-two more conservancies have been established in the northern rangelands, and twenty-eight others intend to follow suit. But harbouring rhinos is expensive. The security has to be tight, and capable of fighting fire with fire. Ol Pejeta itself has a hard core of thirty-two SAS-trained police reservists to back up the daily ranger patrols, which themselves are costly. Richard Vigne calculates that rhinos double or even triple the expense of managing wildlife.

As it is the rhino that brought me here, so by extension it is the rhino that gives a class of neatly uniformed African school-children the chance of a good laugh. It is not every day that a sun-reddened, white-bearded Englishman in dust-covered shorts is brought before them by the headmaster. As I struggle to explain my interest, they find my questions as hilarious as my appearance. Why on earth do I want to know about their exam results? What's it got to do with rhinos? Kenyan school-children are in every way remarkable. All are bilingual in Swahili and English, and most speak a tribal language too. In all, Kenya has sixty-nine spoken languages, though classroom teaching is in English. I bumble away, trying with increasing hopelessness to explain why I have come, and the teenage grins grow ever wider.

Like most schools, Endana Secondary stands at the centre of its catchment. But its catchment is not a town or city with definable streets and communities but a vast stretch of African wilderness. When asked to define it, Ol Pejeta's Community Programme Manager Paul Leringato extends an arm and makes a 360-degree sweep of the horizon. Paul is tall, elegantly dressed, proud of his achievements but no waster of words, and so softly spoken that I have struggled to hear him on our long drive to the school. We have come way beyond the conservancy's bound-aries, past some Maasai living in mud-walled shanties and then juddering across a camel-coloured landscape of pluming dust (which the rains will turn to liquid mud). Herds of sheep and goats, apocalyptically thin, wander far and wide in their day-long search for something to nibble. The distances seem huge, and yet this is the way the children come to school, and there are no bus-routes on the plains. This is why it has a dormitory – a boarding school for village children on the equator! After an hour or so we have turned in through a gate, then bucked and yawed past a well-stamped patch of earth which rickety goalposts identify as a football pitch, to reach a huddle of single-storey breeze-block buildings with corrugated roofs.

The headmaster is Adam Elmoge. He tells me his school has six classrooms, nine teachers and 224 students organised in five streams. They range in age from around fourteen to nineteen-ish, but the classes are not as rigidly age-structured as they are in other parts of the world. Primary education in Kenya is free, but there is no fixed age at which children must report to school. A couple of days later at a chimpanzee sanctuary I will meet a coach-load of primary schoolchildren in the widest imaginable range of sizes. In fact, I hear them before I see them. They are shrieking with pretended terror as an irascible male chimp pelts them with stones from behind the wire. Some of the boys look like men, but they wear their sharply pressed grey uniform shorts with every appearance of pride. It would be a strange thing in Europe or America to see primary schoolchildren looking older than their secondary-school cousins, but here it is all part of the miracle.

Secondary education is not free. Day pupils pay 9,500 Kenyan shillings a year (at the time of my visit, equivalent to £72.12 or $114.39); and boarders 23,627 shillings (£178.92 or $284.49). The compulsory uniform – blue shirt, green pullover, brown trousers or skirt – adds another 4,500 shillings. To a Kenyan farmer these sums do not seem as small as they might to a European or an American. It is an expenditure that has to be thought about, especially when the pupil is a girl for whom no future is envisaged beyond the bearing of children. Even when girls do go to school, says Adam Elmoge, their academic careers can be cut short by pregnancy. This is no surprise. In Mozambique I saw teenage girls at school with babies in their arms. Teachers assured me that the infants were younger siblings being cared for while their parents were in the fields. It might have been true, but the frankly lascivious attitudes of polygamous village men to pubescent girls gave me cause to wonder (I met a witch doctor who believably gave his age as eighty, and whose latest wife was fifteen).

The fees at Endana may be daunting to herdsmen, but – albeit for the opposite reason – they are daunting for the headmaster, too. The boarding fees, he explains, barely cover the students' upkeep, particularly when the country's 18 per cent inflation rate is factored in. They live mostly on maize and beans, and don't have enough books. In the context of a miracle, however, such things are minor nuisances. Miracle is the word. In 2008 Endana had twelve pupils. By January 2010 it had 146, and now (March 2012) it has 224 including sixty-two girls. What has made it possible – what built five of the six classrooms and will soon provide a laboratory – is the rhinoceros. Not the rhino alone, I confess, but the whole living bestiary of Ol Pejeta and the cash it earns from visitors. Every day we pass 4x4s and open-top minibuses glinting with optical arsenals ranging from reflective sunglasses to telephoto lenses the size of rocket-launchers. The drivers stop to quiz each other – who has seen what, where? – but it's not like some national parks (or even the birdwatching hot-spots of North Norfolk), where the bush telegraph gathers a throng for anything rare or iconic. The number of beds on the conservancy is limited to 200, so visitors melt into the landscape like specks of dust. Only once, when a pair of lionesses display themselves on a bluff, do we have to share wild animals with other vehicles. But the visitors are a valuable commodity. Each pays a conservation fee ($68 or £42.87 for a day-trip; less for Kenyan residents and students) and each camp or lodge pays a levy for every night a visitor stays. (If you book a holiday, all this will be included in the price.) The result is what we all see framed in our binoculars – teeming wildlife, with some of the densest concentrations of predators ever recorded in Kenya – and what I now see at Endana School. It is well worth being laughed at. It will not be long before some child of this dusty plain wins a place at university.

The Ol Pejeta Conservancy is 'not for profit' only in the economic sense. The profits are everywhere visible, manifested in gains for the communities of southern Laikipia. The school is one example but there are many more. In a sense what I'm writing is a *mea culpa*. When environmental journalism was in its infancy, some of us, the newly converted, were more inclined to sanctimony than to hard analysis. We were too keen on banning things, and provided an uncritical mouthpiece for campaign groups whose rectitude we took for granted. Carried along by their propaganda and by our own altitudinous rhetoric, we saw every issue as a struggle between man and nature. Wrong and right were as clear as night and day. Wherever such conflicts occurred, it was axiomatic that nature should win. Up with newts! Down with horrible humans! It took too long for many of us to realise the scientific and economic illiteracy of our cause – the knee-jerk opposition, for example, to well-designed and environmentally beneficial applications of GM technology or nuclear power. The perpetual doom-mongering that turned conservationists into technophobes and put environmental politics beyond the electoral pale. The hijacking of the environment movement by the political left, the tendency to submit every issue to trial by ideology, has done immeasurable harm. Rather than destroying the arguments of the free-marketeering libertarian right, they have succeeded only in locking themselves into an unwinnable war of propaganda and misinformation. Until they can acknowledge the benefits, as well as the costs, of GM technology and nuclear power, and recognise the costs as well as the benefits of organics and wind-power, then they will go on shooting themselves in the foot.

On my earlier visits to Mozambique I spent much of my time observing a community forest project at N'hambita in the buffer zone of the Gorongosa National Park. I have already described how civil war had stripped the area of trees and wildlife. At N'hambita another not-for-profit company, part-funded by the European Union, was trying to repair the damage. It was doing this by encouraging farmers to plant trees rather than cut them down, and to abandon slash-and-burn in favour of less exhausting and wasteful methods of agriculture. These had nothing to do with mechanisation, agri-chemicals or anything else that would expose subsistence farmers to risk. They were simply encouraged to intermix their traditional crops – sorghum, maize, cashew, rice, bananas – with pigeon peas. This was soil science at its simplest. Pigeon peas are one of the most useful plants in Africa. They provide an edible crop rich in protein and vitamin B, foliage that can be dug in as compost, and roots that feed the soil with nitrogen. This means the soil stays healthy, the farmers get bigger crops and can go on using the same land year after year without hacking new fields out of the forest.

As a further incentive, they were paid to plant new trees. It was a runaway success. The looming corn grown by the pioneers was a powerful encouragement to their neighbours, and the scheme soon spread to involve more than 1,500 farmers cultivating 2,500 fields in several different communities. In forest clearings I saw pot-grown saplings lined up by the hundred, as neat as a Home Counties garden centre, and tidy rows of vegetable plants being trickle-hosed into plumpness. Soon the farmers were able to produce a bit of surplus, which they could sell for cash. The old mud-hut or outdoor sit-on-a-log schools were replaced with proper buildings. A small clinic appeared, able to offer basic

A girl walks to school at N'hambita in Mozambique. The plant is for
the school garden

medicines and beds where women could give birth more safely than on the mud floors of their huts, and where the authority of the witch doctor was decisively challenged. Who would not applaud such enterprise?

The answer was Friends of the Earth. On my third visit I was accompanied by a camera crew making a film for BBC World, in which I gave as enthusiastic an account of the project as the director would allow. Back in London, for the sake of balance, an opportunity had to be given for a representative of FoE to tell the camera why none of this should be happening. The reason, inevitably, was ideological. Behind N'hambita stood a non-profit company that brokered carbon credits. Imperfections of the Kyoto Protocol meant that it could not be part of any official compliance scheme and so had to be voluntary, but it worked in a similar way. Concerned or image-conscious corporations, organisations and individuals could buy credits from the company to offset their carbon output against tree-planting. Quite apart from its benefits to biodiversity and the carbon economy, this seemed a pretty good way to transfer wealth from the northern rich to the African poor. A bush secondary school with a computer on every desk? Not an impossible dream. Here it had already happened.

So why did FoE decry it? Simple. It had a policy of opposition to carbon trading, and so was constitutionally disbarred from acknowledging any benefit derived from it. Equally predictably, I soon found myself being vilified from the opposite wing of the belief-spectrum by a climate-sceptic blog which implied that I was a paid stooge of the company that founded the project. This is a pretty good example of the debased condition into which 'debate' has fallen. The same blog chose to see something sinister in the fact that *The Sunday Times* had asked me to interview Professor Phil Jones, the man at the centre of the

so-called Climategate scandal (when emails leaked from the
Climate Research Unit at the University of East Anglia seemed
to suggest that data had been withheld or fiddled with). As it
happened, the interview had been fixed by arrangement between
the *Sunday Times* news editor and a media consultant hired by
the university. This man, who was unknown to me, was a former
executive of the now defunct *News of the World* who later would
be arrested over allegations of phone hacking (though he was
subsequently released without charge). The blogger purported
to think that this proved an 'illicit relationship' between myself
and the man's PR firm – further reason, if any were needed, to
disbelieve every word I wrote. There was nothing unusual in
this. Anyone who enters the climate-change arena can expect to
be smeared, and compared with others I got off lightly. My
regret is that by over-simplification in the past, and by promoting
a doctrine of absolute truth, for all our good intentions we may
ourselves have contributed to the moral and intellectual implo-
sion that makes such nonsense possible.

For far too long, the natural and human worlds have been
perceived as warring entities whose interests are irreconcilable.
In their different ways, both N'hambita and Ol Pejeta have
shown this to be false. Ol Pejeta's chief executive Richard Vigne
puts it like this: 'We are moving away from the idea of fortress
conservation that takes place behind fences in the absence of
any other human activity. We're saying that if conservation is
going to continue on a landscape scale, then we're going to
have to accommodate people and their activities in some way
or another. We are opening up more opportunities for conservation by overcoming the mindset that you can only do it where
there are no humans.' This is what's in my mind when headmaster Elmoge describes the Endana curriculum. All pupils at
the school are taught English, Swahili, maths, chemistry and

religion, and may choose between biology and physics, history and geography, business studies and agriculture.

'Crop science,' says Adam Elmoge when I ask what is taught in the agriculture classes. On the evidence of what I've seen, this could sound like a joke – lessons in cake-making in a land with no bread. But it could not be more apposite. The lights that begin to glow as twilight falls across the lower slopes of Mount Kenya are from nurseries growing vegetables that will find their way to British supermarkets. I am always irritated by green evangelists who bang on about 'localism' and 'food miles', as if there were something un-green about eating African beans. If 'green' means a co-operative and sustainable sharing of the world's resources, then what could be greener than supporting Kenyan farmworkers?

Of course this is not the issue at Ol Pejeta. The priority here is subsistence, not exports, but this doesn't alter the message. As sons and daughters are being taught in the classroom, so their parents are learning in the fields. At the conservancy head-quarters I meet Josphat Kiama, Ol Pejeta's Agricultural Extension Officer. He is young, energetic and persuasive, at once idealistic and pragmatic, driven by outcomes rather than ideologies. The keyword is *productivity*. This doesn't mean telling traditional farmers to radically change their ways. As at N'hambita, it means showing them how to make the old ways work more efficiently. Example: it takes eight men four days to weed an acre by hand. With Roundup weedkiller one man can do it in an hour. This is much cheaper, especially when farmers combine to bulk-buy the weedkiller (and let the organo-fascists fulminate as they may). The same is true of seed and fertiliser. This is about life, not lifestyle. And it is about cooperation. Example: one farmer owns a drilling machine and lets others borrow it for the price of the diesel.

Josphat is high on practicality, low on cant. 'Sustainability' is not some gaseous extrusion of environmental Newspeak, *pace* Gro Harlem Brundtland, but the very lifeline by which families cling to the soil. The landscape asks brutal questions; Josphat provides uncomplicated answers. To keep the soil healthy he prescribes a simple crop rotation – maize, beans, potatoes – and mulching to retain the moisture. Soil disturbance is minimised by bio-friendly 'no-till' techniques that require no ploughing. This ensures the survival of worms and micro-organisms, and keeps carbon locked in the earth. The ground is disturbed only as far as it is necessary to implant the seed, which is watered by a drip irrigation system that delivers what the crop needs and not a cupful more. It uses only a sixth of the water consumed by haphazard sloshing. The no-till method is also parsimonious with fuel. Petrol consumption per acre is down to half a litre a week – previously it was six.

Livestock, too, is being improved by selective breeding. The arrival of Dorper sheep from South Africa means that animals now reach market weight in six months instead of three years. This may not delight the European welfare lobby but it's good news in equatorial Africa. For milk, Josphat favours the goat – much more likely than a cow to withstand drought. He also encourages farmers to grow and store hay rather than expect their animals to survive on the desiccated scraps that nature provides in the dry season. No one has to take his word for it. He can talk the hind legs off a giraffe, and reduce to jelly the writing arm of a visiting note-taker (having made a point, he will not move on until I have written it down), but he knows it's example that counts, not explication. As it was at N'hambita, so it is here. It is the early-adopters, the emboldened pioneers enjoying their heavier crops and healthier animals, who are the best recruiters. So it is that in a single season the human benefit from Ol Pejeta passes from classroom into field.

Back in the conservancy, still thinking about the mole and what it represents, I recall another childhood visit to London Zoo. In a shady corner, away from all the big attractions – which in those days included elephants, rhinos, big cats, bears and wolves – a crowd was gathering. Cameras were clicking. Parents were shouting for their children (including me) to come and look. *Ooh*, we went, and, *aah*! The object of our admiration paused for a moment – I would guess now in bewildered fright, but at the time I imagined it was playing to the gallery – before vanishing suddenly skywards up a tree. It was an ordinary grey squirrel, a mundane but suddenly unignorable anomaly in the company of lions. On the plains of Ol Pejeta I experience a similar moment of disconnection. The circumstances are different, and there is a vastly different scale of magnitude, but the sense of displaced ordinariness, of the mundane made exotic, is weirdly alike. Often during our game drives we would see distant clusters of tour vehicles, flashing like diamonds in the sun. It was not the plains game that drew them. Not big cats. Not rhinos, giraffes or elephants. Not even anything small and cute like a squirrel.

It was cows. Ranch-scale herds of domestic cattle, sharing the land with lions. These are not, it must be admitted, the Herefords and Friesian-Holsteins of the English lowlands. They are spectacular Ankole, whose curved horns can reach 8 feet from tip to tip, and humped Borans like the sacred cattle of India. Both are exotic to western eyes, but they are cattle all the same, with the same needs and vulnerabilities as any herd in the Cotswolds, but here with the added spice of the world's top predators – a somewhat more exciting risk than badger-borne bovine TB. In a very direct way, though unconnected with this apparent supply of easy meat, the lions are beneficiaries of the cattle. So too are the impala, the zebras, the rhinos, giraffes, leopards and hyenas . . . So is everything that lives here, from the raptors overhead to the subterranean *confrères* of the elusive golden mole. Even the schoolchildren in their classrooms and laboratories, at least in part, owe their improving exam grades to Ol Pejeta beef.

This duality is not something you will see in Kenya's National Parks, where a purist philosophy deems ranching and wildlife to be incompatible. Neither is it something you would have seen years ago in the ranch-lands. 'In the old days,' says Richard Vigne, 'the feeling was that if you wanted to succeed as a cattle rancher, one thing you had to do was eliminate wildlife from your land. And that's what they did.' So Ol Pejeta is different. It refutes the old idea that a gain for wildlife is a loss for humans, and it recognises that it's not enough simply to rewrite the law in favour of conservation. For millennia, people on this continent lived by hunting. At sea, aboriginal fishermen have been allowed a quota for subsistence, but on land the ban on most traditionally hunted species is absolute. Anyone caught poaching – provided they are not corrupt officials protected by their peers – should be caught and punished. Flesh from wild animals, once just plain 'meat', is now illegal 'bushmeat'. I think yet again of the Chitengo Two, held for killing a warthog. I think, too, of the injustices of English nineteenth-century poaching laws, when poor village men snaring rabbits to feed their families were viciously man-trapped, shot at and transported to penal colonies in Australia. Whatever word you might choose to describe this, it is unlikely to be 'justice'.

Ranching at Ol Pejeta is important for the very same reasons that tourism is. It creates the income the conservancy needs to protect wild animals; it reduces the reliance on charitable dona-tions (though these remain important); it demonstrates to other landowners the viability of managing their hectares for wildlife, and it gives the local communities something in return for their cooperation. That 'something' is substantial. It includes road improvements and piped fresh water as well as agricultural improvements, support for a hospital, three health clinics and the schools. It's not just Endana. In all, the conservancy supports twenty local schools. On average at any time, forty secondary school students will be maintained on full-time bursaries (awarded in consultation with local leaders and community groups), and some 250 on part-time bursaries given in periods of hardship such as drought, when resources are stretched. There are gifts of books, desks, chairs, laboratory equipment, water tanks and computers. Five schools will be given a couple of cows each, so that they can be self-sufficient in biogas and milk. The camps within the conservancy also play their part – Kicheche sponsors bursaries and has helped to set up a chicken project providing eggs for the local orphanage, to which it also gives blankets. Most importantly Ol Pejeta creates jobs – 700 of them altogether, of which 80 per cent are filled by local people who otherwise would have no prospect of employment. All that would be open to them, says Richard Vigne, would be 'scratching around on sub-economic plots of land'.

In a way this is a dangerous argument. Emphasising the community benefits of projects such as Ol Pejeta is to invite the conclusion that conservation alone is not enough. It is not easy to make a business case for saving wildlife – economists can always find more intensive ways of using land – but Ol Pejeta strikes me as the model answer. Philosophers such as Peter Singer have argued the case for universal rights, extending to all species the utilitarian principle that the only ethical standard by which behaviour can be judged is, as Jeremy

Bentham put it, 'the greatest happiness of the greatest number'. Singer and others specifically reject 'speciesism' – the inherent belief that *Homo sapiens* has a special entitlement and should enjoy rights denied to others. As they see it, all exploitation of animals by humans should cease. No rhino horn, no ivory, no food animals, no hunting, no pets. The case is morally perfect, smooth as an egg, but no likelier than an egg to survive hard impacts with an imperfect reality. Charred animal bones provide some of the earliest evidence of human settlement, and there is very little in all the millennia to encourage the view that every 'possessor of a life' is equal in the eyes of the world.

Nature is all about power, and in the Anthropocene the power is all ours. Viruses apart, there is literally nothing we can't find a way to kill. Somewhere within us, however, is something felt rather than reasoned, a well of sentiment that persuades even the simplest mind to travel at least part-way with Jeremy Bentham. Animals earned their right to humane treatment, Bentham argued, not through their capacity for reason but through their ability to suffer. From whatever source it may come – evolution, religion, education – we have a revulsion to cruelty, which we commonly describe as 'inhuman'. This is not quite universal. Disregard for others' feelings – human as well as animal – is likeliest where survival is hardest. *In extremis* we would eat our neighbours. But most of us are not *in extremis*. Tim Flannery speaks of a 'commonwealth of virtue' in which people of all cultural, racial and economic backgrounds recognise each other's goodness. In his book *Here on Earth: a new beginning*, he writes movingly of the 'natural magic' of an encounter, in a remote part of New Guinea, with local people whose ancestors and his own had parted ways not long after the dawn of civilisation. 'Yet when we met, after fifty millennia of separation, I understood immediately the meaning of the shy smile on the face of the young boy looking at me, and he understood my motion for him to step closer to better observe what I was doing.'

Charles Darwin left us with a question to ponder. Does evolution tend towards increasingly stable ecosystems that bind together every living thing in a Gaian mesh of symbiotic relationships? Or must it lead catastrophically to an all-powerful 'Genghis Khan species' whose unquenchable greed will exhaust the planet? Flannery inclines optimistically to the former, trusting in the capacity of human genius to take us beyond the 'civilised imbecility' of the early Anthropocene. There are times and places where it is easier to be carried along by his optimism. One of these is Ol Pejeta, where, somewhat over-fortified by a good supper and a generous quantity of South African red wine, I scribble the bare bones of this paragraph by torchlight on the veranda of my bungalow-tent. The night sounds would be frightening if, like Roualeyn Gordon-Cumming, I were bivouacked in the open. Birds over the water – cranes, I think – are whooping like drunken oboists in an orchestra of improvisation. There are deep mammalian grunts, coughs, splashings and a long, choking wail that must surely signal the letting of blood. The darkness seethes with teeth, claws and padded feet. Watching a raft of pelicans sailing on the moonlight, I submit to the rush of sentiment. In the end it's not about philosophy or economics. It's about the way we respond to the world around us. Either we feel enriched by the presence of other species, and will honour their right to exist, or we don't.

Next day I visit the cattle. One of the most encouraging features of Ol Pejeta is the absence from the management team of white faces. Most of the specialists – in conservation, agriculture, education, community development, wildlife management and research – are black Kenyans. The only conspicuous whites are the chief executive Richard Vigne, the security adviser Batian Craig, both of whom are Kenyan-born, and the livestock manager Giles Prettejohn, a fourth-generation Kenyan cattle rancher. To find Prettejohn, Andrew has to drive us – Richard Lamprey and me – far across the plain to a cluster of farm buildings. From a distance it could be mistaken for a transport depot in the Fens. Behind one of the buildings we meet a throng of enormous Marabou storks, whose popular name, 'undertaker birds', is an apt description both of their sinister aspect and of their morbid habits. I ask Andrew what has attracted them. He points. The building is Ol Pejeta's slaughterhouse, and they are gathered by the drain.

Giles Prettejohn is a bluff, weather-beaten man of the outdoors whose impeccable Home Counties accent would blend him into any gathering of gentlemen farmers at an agricultural show in England. With his boots on the equator, however, he is somewhat more conspicuous. What could be more unlikely, in an area dedicated to African big game, than a beef farmer? Surely, I say, you might as well try to raise seabass in a shark tank as graze cattle in full sight of lions. But Giles is no fantasist. I don't have the statistics to prove it, but I suspect he is one of the biggest beef producers in Kenya. I *do* have stats to show that he runs the biggest herd of pure-bred Boran cattle in the world. Across Ol Pejeta's plains are spread 6,000 – yes, *6,000* – of them, including 2,000 breeding cows. It is a fact that the number of lions in the conservancy has quadrupled in the last six years and has already exceeded their theoretical carrying capacity, but this has little to do with preying on Giles's cattle. Small wonder that visitors come from all over Africa to see in action this remarkable twinning of opposites, which in most parts of the continent, including South Africa, would be anathema for reasons both ideological (the natural sanctity of wildlife reserves) and commonsensical (the exposure of cows to predators). Everything here is counter-intuitive. The cattle not only thrive and make good money, but they have turned out to be powerful allies of the wildlife.

The Borans with their fatty humps have a primitive look about them. This is not misleading, for they are antiquity on the hoof. Giles, who knows as much about the breed as any man alive, tells me that they arrived in Africa from India some 2,000 years ago, and were kept by the tribe of the same name – the Boran of north-east Kenya and southern Ethiopia. Given the hostility of the equatorial environment – scrappy forage, drought, disease – the animals were necessarily small and hardy. Recently they have been beefed up by cross-breeding – a bit of shorthorn here, a bit of Hereford there – but not so much as to change their appearance or character, and certainly not their attraction for predators.

As lions hunt by stealth and not speed, much of their killing is done in the dark. Compared to the ill-tempered, muscular aggression of buffalo and the agile speed of gazelles, the docile, slow-moving cattle look like the world's biggest free picnic. But Giles has a simple and ingenious answer to the Borans' nocturnal vulnerability. It also explains some puzzling brown circular patches which I noticed from the air. He takes me first to see a 500-strong herd of pregnant cows, grazing freely on the plain and watched over by herdsmen. Then he shows me the magic ingredient.

It is called a *boma*. In principle it's like a native *kraal*, a circular enclosure within which cattle are kept secure at night, but unlike a *kraal* it doesn't involve mud walls or palisades, and – this is the essential difference – it is *movable*. Sections of tubular aluminium fencing are driven into the ground and pinned together in a circle, enclosing an area about the size of a tennis court. Into this small area at night, 500 cattle will be packed so tightly that they cannot move. Their immobility is essential. It means that if lions visit the *boma* at night, the Borans are unable to stampede. If they did, the *boma* would be knocked over and the lions would have an orgy. It also mimics the Borans' natural behaviour. At night on open pasture, for comfort and protection, they will instinctively form a huddle. The fencing is not proof against incursion by lions – they could jump over it if they wanted to – but their innate wariness of buffaloes inhibits them from leaping into a mass of cattle. Even if they did make a kill within the *boma*, they would be unable to escape over the fence with it.

This does not mean an end to all predation. Thwarted at night, the lions turn their attention to daytime when the sheer weight of numbers means that cattle and cats are bound to meet at some point. Giles loses about sixty-five animals a year, which from a 6,000-strong herd he reckons is acceptable. If a particular lion makes too much of a nuisance of itself, then it can be darted and moved away. The cattle themselves are constantly shifting to new grazing, and the *bomas* are moved every three weeks. Something magical then happens. You get a clue just by looking at an empty *boma*. The ground inside is entirely bare of grass, not a blade anywhere, just a deep dark churn of well-watered manure. You can smell it, the distilled and well-ripened essence of cow. Insects ping off your skin like hail. Pumped full of nitrates and phosphates, the ground is a brimming reservoir of pent-up energy. At the first slick of rain, due any day now, it will explode into life. And here's the magic. The new growth will not be the same rank old bamboo grass that was there before. It will be of a wholly different kind. In a sudden flood of emerald green there will appear star grass (of the species *Cynodon*), which is much more palatable and nutritious for the animals. Why this happens, no one seems to know. What matters is that it *does* happen, and that it is yet another link in Ol Pejeta's virtuous circle. What's good for the Borans is good for the plains game that moves in behind them. What's good for the plains game is good for the predators. What's good for the predators is good for the tourists and hence for everything that depends on their money.

Nor is that the only gain. Wild buffalo carry East Coast fever, a tick-borne disease which even an indigenous breed like the Borans cannot survive. For this reason the cattle have to be sprayed regularly with insecticide, which has not only the desired effect of keeping the cattle healthy but also prevents the transfer of ticks to other species. All this is well and good, but there is more to the story than its incidental benefits. Giles is first and foremost a stock farmer, and it matters that the herd makes sense economically. Back at the farm he shows me the slaughterhouse. The day's work is over and a thicket of cream-coloured carcasses hangs from hooks, giving off that odour of freshly killed meat that you either love or hate. Each side shows a layer of fat and well-marbled flesh, just the way a good butcher likes it. Every week Giles sends between fifty and sixty carcasses to the Nairobi meat market, where they fetch premium prices. Because I've never seen one butchered before, I'm interested in the hump. At 50 per cent fat it is the richest cut on the carcass, particularly prized, Giles tells me, by Nairobi's Asian community, who like to boil and slice it very thin. My one regret about the sumptuous meals at Kicheche Camp is that hump never appears on the menu (though I do enjoy a plateful of stir-fried Boran steak).

In a hungry pan-African market, none of this passes unnoticed. Demand is high, especially from South Africa, which imports Boran embryos direct from Ol Pejeta (embryos only, because it is illegal to move live animals across national borders). Though ranching in South Africa is kept separate from wildlife management, visitors from elsewhere – Zimbabwe, Botswana, Namibia – are showing interest in the full Ol Pejeta model. For Fauna & Flora International it is a spectacularly powerful testament to the viability of what might have seemed a most improbable kind of investment. It's not quite the lion lying down with the lamb, but standing up with the Boran runs it close. Yet you never forget where you are. The wilderness may be managed, but it is wilderness all the same. In all the many miles that Andrew drives me, I never see an empty landscape. When the time comes for Richard Lamprey to depart, we find his Cessna is sharing the airstrip with zebras, warthogs, Grant's gazelles, a Jackson's hartebeest and a giraffe. Wherever flyers in Africa gather, there is gallows talk – remembrances of this old acquaintance or that coming down in the bush, or of a pilot of long experience finally copping the Big One, often as a result of wildlife on the strip. Richard, who has survived his share of incidents, takes his time checking the aircraft over. Not even the fuel gauge is taken on trust – he checks the level with a dipstick. It brings home to me how remote, how wild, this place really is. Take away the aeroplane and the Land Cruiser, and there's nothing here that would strike a Gordon-Cumming or a Selous as remotely untypical of the nineteenth century. In all important respects, and despite all the care lavished upon it, this is quintessential Africa, exactly as nature meant it to be, and it's humans who must adapt. Flying from here demands a technique not taught to pilots at Heathrow. Taxiing is as much about clearing animals from the track as it is about lining up for take-off. Then there can be no hanging about. Quick as you can, you have to turn and make your run before the moment passes. Only the giraffe doesn't move. It stands at the end of the strip, head like a windsock. As much as the mountains and trees, it seems to belong to the solid fabric of the place, part of the landscape. It doesn't flinch even as the Cessna, suddenly tiny, flashes over its head. The plane climbs rapidly, turns south and dissolves into silence.

That should have been the end of the chapter. During the writing of it, however, I took a short holiday in Istanbul. We did all the usual things – a trip along the Bosporus, visits to the Blue Mosque, Hagia Sophia, archaeology museum, spice market, the Topkapi Palace. In a far corner of the Topkapi stands the Baghdad Pavilion, built in the 1640s and used during the Ottoman Empire for Cabinet meetings. Descriptions of the decorative art at Topkapi tend to overwork the word 'exquisite'. But what else can you say? Inside the Baghdad Pavilion you can't see the walls for all the photographers taking close-ups of the detail. There is a famous classical fireplace, framed by ceramic tiles depicting birds; niches along the walls decorated with Iznik tiles older than the building; an ornate silver brazier given by Louis XIV. The whole thing stands as a monument to the absolute power of a Sultanate that was never more than a handclap away from anything it wanted. Human lives were cheap; animal lives worth only what could be made from their body parts. Truly, this often was exquisite. The pavilion's dome is lined with floral patterns made from gazelle leather. Window shutters and cabinets are inlaid with mother-of-pearl, turtleshell and ivory. But this was all so long ago, long before it was understood that animals were a finite resource, that there is no sense of outrage that gazelles, elephants and turtles had to die for a decorator's whim. You can't blame a sultan for being a man of his time. The world is different now. No monarch of *this* time would indulge such lethal ostentation.

But then we go back to our hotel and switch on the news. There is a royal scandal. King Juan Carlos of Spain – honorary president of the Spanish WWF – has been caught elephant-shooting in Botswana, and has made a grovelling apology to his people. *I am very sorry. I made a mistake. It won't happen again.* The mistake, it turns out, was to have tripped in a hunting lodge and broken his hip. News of the mishap dominated the Spanish headlines for days, arousing the kind of passionate anger for which that country is unrivalled. Politicians across the spectrum at last had something they could agree about – the king would have to eat crow. When he did, however, it was not for killing elephants that he apologised. It was for having taken such an expensive holiday at a time when his subjects were suffering hardship and the economy had holes in its shoes. Killing elephants was wrong because poor people could not afford it.

Worse was to follow. Back in England I find an email from Brian Jackman telling me that Ol Pejeta's near neighbour, the Lewa conservancy (a favourite, apparently, of young British royals, where William proposed to Kate), has lost a pregnant black rhino, shot by poachers despite one of the tightest security operations in Africa. He reminds me that more African ivory was intercepted in 2011 than in any year of the last two decades. Cameroon's Bouba N'Djida reserve alone had lost more than 450 elephants. In South Africa, the number of rhinos lost since the beginning of the year (I am writing this in mid April 2012) had already reached 170. I have enjoyed Ol Pejeta. It has been one of those rare and exalted things that might be called an experience of a lifetime. Briefly I have been intoxicated by optimism. I have seen how life for animals

can mean livelihoods for people; how good people can make
a difference. But there is no hiding from the bad. Brian quotes
a friend of his, centrally involved in the conservation of rhinos,
who speaks unemotionally of 'an unfolding disaster for Africa'.
I know now why the mole matters. Conservationists in soapbox
mode tend too easily to slip into eco-jargon, as if words like
biomass, sustainability and *biodiversity* carried some unques-
tionable authority, like edicts from the Vatican. Tim Smit,
founder of the Eden Project and as much a champion of plain
speaking as he is of the environment, tells me of a survey at
the Natural History Museum which revealed that 85 per cent
of its visitors didn't know what biodiversity meant. 'That tells
you two things,' he says. 'It tells you first of all, *Ouch!*, in
terms of our education system. But it also tells you what a
bunch of arrogant tosspots we all are, using a phrase that 85
per cent of people don't understand, when we could say *variety
of life*.'

Absolutely this is what matters. We understand, more or
less, that varieties of life connect to each other in complex webs
of inter-dependent relationships that we call ecosystems. The
survival of a large predator at the top of the system depends
on linkages that reach all the way down to micro-organisms at
the bottom. You can't have one without all the others. There
is some flexibility. A species may die or be expelled and another
will move into its niche. But there will come a breaking point.
The biologist Paul Ehrlich likens it to the piecemeal disassembly
of an aircraft. You can go on removing some of its tiniest
components – rivets from the wings, say – and for a while it
will go on flying. One day, however, you will remove one too
many and the system will crash. Conservation is about saving
the rivets; fixing the ones that are working loose; catching some
even as they fall. Maybe the Somali golden mole is still part

of a working mechanism; maybe it is a missing rivet. If it *has* fallen, then the world is in some small way poorer for its loss. As a proxy for every other unseen or unheard-of subterranean toiler, it deserves, at the very least, a pilgrimage and a decent obituary.

CHAPTER TEN

Unpronounceable Teeth

The pilgrimage is not going to be easy. My naivety now is being worn down like topsoil in the wind, exposing hard little spikes of scepticism. I have to acknowledge that I've been kept back from the search by a growing expectation that it will end in failure. The world expert on golden moles, Gary Bronner, has told me that the University of Florence could not produce Professor Simonetta's barn-owl pellet for his inspection. Now I can delay no longer. The time has come for me to beat my head on the same door. My grasp of Italian does not extend much beyond *Vitello tonnato*, so it takes me some time to click my way around the *Università degli Studi di Firenze* website and find the name of its communications supremo, Silvia D'Addario. I compose a carefully worded email in English, explaining my business and framing my questions – is Professor Simonetta still alive? Does the zoology department have the holotype? – and hit the Send button. My expectations, however, are low. Institutional bureaucracies are difficult to penetrate, and some recent encounters have wrestled me almost to a standstill. In particular, I have been sandbagged by the Natural History Museum, which suddenly wants money to answer my questions. I had asked: What was the origin of the stuffed giant golden mole, *Calcochloris trevelyani*, in the museum's mammals gallery?

Were any specimens in the collection shot by Roualeyn Gordon-Cumming or Frederick Selous? Not vital questions, to be sure, but historical details of particular interest to anyone fascinated, as I am, by this rare collection of bygones.

The answer to the first, I later realised, I had already found for myself (see Chapter Three), but I still wanted to know about Selous and Gordon-Cumming. The reply came from a department previously unknown to me, NHM Research Consulting, which explained that (unlike newspaper enquiries, which are answered rapidly and *gratis*) the museum was 'obliged to charge' for helping authors. Answering my questions apparently would require four hours' work, charged at £95 for the first hour and £75 for every hour or part-hour thereafter, so I could expect a bill of £320, which VAT would stretch to £384. This saddened me not just because it makes fact-checking unaffordable but, more importantly, for the hole it exposes in the museum's own recorded history. The statue of Frederick Courteney Selous stands, rifle in hand, in the Central Hall alongside Charles Darwin, Thomas Henry Huxley and Richard Owen. And yet, extraordinarily, the museum has no accessible record of the surviving exhibits it owes to him. To grind salt into the wound, an internal memo attached to NHM Research Consulting's email even manages to misspell his name, 'Salou'. Somehow I manage not to weep.

But the gloom lifts. Next day I get a friendly email from Florence. Great news! The professor lives! Silvia D'Addario has passed my message on to him, and is not asking to be paid. Like a teenager on Valentine's Day I haunt my inbox waiting for the professor's reply. Realistically, of course, I know it could take weeks (I had to wait nearly a month for the helpful Gary Bronner), or it might not come at all. Meanwhile, confidence plummeting, I submit finally to the realisation – which has been gradually dawning on me – that my knowledge of zoology is more appropriate to the five-year-old who fell in love with Brumas than to the white-haired author now scratching his beard. Viewing a chimpanzee, I can easily enough believe in the shared ancestry of *Homo sapiens* and *Pan troglodytes*. But how can whales be related, as I am assured they are, to hippopotamus and deer? I now find myself stuck even for an adequate understanding of what constitutes a *species*, never mind a *genus*. Reassuringly, I find I am not alone. Bigger brains than mine, including Aristotle's and Darwin's, have been gnawing at these questions for centuries, and certainties are as elusive as the quagga.

I find I am in dire need of expertise. With the Natural History Museum now out of my price range, I try the Royal Society, which kindly suggests the Science Media Centre, which couldn't sound more discouraging if it tried. *Their* job, a woman patiently explains, is to help journalists with hot scientific news, not chase philosophical phantoms with writers who can't even explain their books properly. Nor can she think offhand of any available expert on taxonomy – rarely the subject of hot scientific news – but she says she'll ask around. And, glory be, she does. Within a few hours I get a call from the Centre's mental health expert, Joe Milton, who (I thank the gods of serendipity) turns out to have a Ph.D. in taxonomy. Would I like to see his thesis, which – serendipity alpha plus! – addresses the very issues I am interested in? Precisely eight minutes after the email from Silvia D'Addario, Joe's paper drops into my inbox.

'Phylogenetics' is not a word to quicken the pulse. It is exactly the kind of philological construct that would offend the plain-speaking Tim Smit, a lofty palisade against common understanding. And yet it offers the most graphic account of what we all are, where we've come from and where we might be going. It's all about family trees – 'reconstructing evolutionary history', as Joe Milton puts it. 'Phylogenetic trees' are those familiar branching diagrams that show relationships between species and track them back to common ancestors. It's as simple, and as complex, as that. Somewhere, way back in the primal fog, the golden mole and I were one.

The naming of species is as old as language. Early hominids will not have taken long to realise that the animals they saw were not infinite in their variety, though they were many, but were organised like themselves into discrete groups of approximate lookalikes, all reproducing themselves in their own image. For each of these they would have had a word or a sign. There were animals you could eat, and animals you could be eaten by. Nothing was more important than knowing which was which. As the millennia rolled by, animals were a decisive influence on human culture, subjects of graphic and culinary art, symbols of economic power and objects of worship. Animal imagery was fundamental to the ancient cultures of Egypt and Greece, and has been a rich vein of metaphor ever since. To suggest everything from undying love to bitter contempt, we liken each other to apes, asses, badgers, bats, bears, bees, cats, chicks, cockroaches, cows, dogs, donkeys, eagles, elephants, fleas, foxes, hens, hyenas, jackals, lemmings, lions, mice, monkeys, moths, ostriches, owls, pigs, puppies, rats, skunks, tigers, turkeys, whales, wolves, and even once in a while to moles. The distinction between species is so basic to the way we look at life that, outside the community of natural science, we seldom give it a critical thought. *That's a horse. That's a cow. That's a Rafinesque's big-eared bat.* All are immutable products of Creation/evolution, which come indelibly labelled and neatly parcelled in family groups. Or so I had always thought.

'Taxonomy' (not to be confused with taxidermy) is the science of labelling and organising these groups. It has a long history. Joe's thesis traces it back to Pliny, Aristotle and Theophrastus, all of whom devised formal classifications of living things. The system of plant classification developed by Theophrastus in the third century BC, which earned him the sobriquet 'father of botany', remained in use until the eighteenth century. Only then was it superseded by the work of Carl von Linné – the *soi-disant* Carolus Linnaeus – whose *Systema Naturae*, first published in 1735, marked the beginning of modern taxonomy. Linnaeus, with commendable but naive over-ambition, believed he could record absolutely everything that flew, swam, walked or crawled on earth. He was a man of deep religious conviction who believed his task was to 'reveal the Divine Order of God's Creation'. It might be no coincidence that it chimes so closely with the verse in Genesis (2: 19) in which Adam was commanded to do likewise:

> And out of the ground the LORD God formed every beast of the field, and every fowl of the air; and brought them unto Adam to see what he would call them: and whatsoever Adam called every living creature, that was the name thereof.

Adam never finished the job; nor did Linnaeus; nor has anyone since, and I can't find anyone who believes there is any realistic chance of the 'Divine Order' ever being fully revealed to Adam or anyone else. My own luck, however, continues to improve. Only two days after Joe Milton's call, to my great and happy surprise I find myself back, *gratis*, in the Natural History Museum being shown specimens collected by Frederick Selous. These are not mammals, however, but insects. They are hide beetles (that's 'hide' in the sense of animal skins), busy operatives in nature's clean-up squad, which feed on the skins of dead animals. For collectors like Selous and Gordon-Cumming they must have been an expensive nuisance. They are about the size and colour of olive pips and, to anyone but an entomologist, about as interesting to look at. But never mind. The man who displays them is keen to the point of obsession. Maxwell Barclay is a premier-league entomologist, in charge of the museum's enormous beetle collection. To spend a few minutes in his domain is to recall the Scottish biologist J. B. S. Haldane's famous aphorism: 'The Creator, if He exists, has a special preference for beetles.' There is ample evidence, too, of the prescience of Haldane's other oft-quoted remark: 'Now, my own suspicion is that the universe is not only queerer than we suppose, but queerer than we *can* suppose . . . I suspect that there are more things in heaven and earth than are dreamed of, or can be dreamed of . . .'

I had asked Joe to recommend an expert who would have the patience to talk me through the basic rules of taxonomy and perhaps make the universe seem just a little less queer. Max Barclay has generously volunteered. Yet again I ascend the familiar staircase past the sculpted giants – Darwin, Selous, Owen, Huxley. A Chinese girl is photographing Darwin, crouched down and shooting upwards past his huge marble boot. No one is paying any attention to his neighbours. The entomology department is a low cluttered space crammed like a wasps' nest high under the eaves. Max Barclay is a brilliant lecturer and I form an attentive audience of one. He confirms what I have already suspected. Had I been writing a hundred years ago, the issue would have been a whole lot simpler. Then there was a broad understanding, though no clear definition, of what was meant by 'species', and animals were grouped into 'families' on the simple premise of how closely they resembled each other. It's a whole lot more difficult now.

Aristotle in the fourth century BC had some notion that appearance wasn't everything – he was able to conclude, for example, that a whale was not a fish – but even Linnaeus, who had no idea about evolution, was noting similarity rather than relatedness. As Max points out, similarity is a fairly good proxy for relatedness – we all tend to resemble our families – but it is not the whole story. Darwin began to realise this, and believed the system of classification should be more like genealogy. Max explains how this works in practice: 'Instead of a series of separate boxes into which things were thrown because they looked alike, he was working up a system that looked like a family tree, where you could trace relationships between living and fossil groups, and between different living groups.' It is not a term that Darwin himself would have used, but this is where phylogenetics began.

It didn't put an end to the difficulties, though. Even now, no one has come up with a definition of 'species' that works for all living things, though for higher organisms such as mammals there is an accepted rule of thumb, first proposed by the twentieth-century German evolutionary biologist Ernst Mayr. A species, he said, was not just a group of plants or animals that looked alike. It was, literally, a matter of breeding. According to his 'biological species concept', any two organisms that can breed together and produce fertile offspring are members of the same species. If they cannot breed, or can produce only infertile offspring (as in, say, a mule from a union of horse and donkey, or liger from lion and tiger), then they are of different species.

This was a significant refinement of Darwin's theory of evolution. If a population was isolated from others of its kind, then over many generations it could diverge so widely from its cousins that it was no longer capable of breeding with them. This might happen because of different diets or feeding patterns, mating behaviour or natural selection. The next theoretical tweak was by Mayr's compatriot Willi Hennig. It is now that another polysyllabic brute, 'phylogenetic systematics', enters the lexicon, though only to be supplanted by yet another, 'cladistics', meaning the study of 'clades'. Just in typing the word I fear the ire of Tim Smit; another barrier to non-speakers of zoologese. Yet I find it is quite a useful one to hang on to – a good theoretical handhold. A clade is a single branch of the evolutionary tree comprising a species and all its descendants. The tree metaphor here works perfectly. Clearly the length of a clade will vary, depending on how far back up the tree you climb. Long clades may include many shorter ones – the side-branches and twigs that split from the main stem. Thus a clade may contain just a few species, or many thousands.

Hennig's big idea was a scoring system designed to show just how closely related various species, or groups of species, actually were. It is not just a matter of counting points of similarity – that might lead you to put whales in the same clade as fish – but rather of deciding which characteristics are the ones that matter. Crudely simplified, it means distinguishing between characteristics in their most primitive, ancestral state (*plesiomorphous*, if you want the technical term) from the derived or advanced (*apomorphous*) states that have been passed down the line. It is these derived characteristics which contain the 'phylogenetic signals' and mark out family relationships. Simplifying further, it allows scientists to distinguish between species that are genuinely related, and those which have developed similarities – long probing snouts, powerful digging claws, dense waterproof coats, for example – simply through sharing similar environments. As an entomologist, Max takes his example from the world he knows best. 'If you look at swimming insects,' he says, 'they all look very similar. They all have fringes of bristles along the legs, they all have some kind of water-repelling structures, and they all have some kind of air-carrying structures.' Usually, however, they have developed these characteristics independently and have not inherited them from common ancestors. *Lost* characteristics can be just as misleading. 'Under a Linnaean system,' says Max, 'flightless birds might all be put together as a group, whereas it may be that many birds that have appeared on oceanic islands have lost the ability to fly without in any way being related to one another.'

All this bears directly on nomenclature. At the time Linnaeus was devising his binomial system, most educated people in western societies would have known a smidgen of Latin. In every way, it made good sense. Having a shared academic language eased the problem of international communication. Everyone, or everyone who needed to, knew that *Homo* meant 'man', and *sapiens* meant 'wise'. The *sapiens* bit might be arguable, but Latin speakers knew their species from their genus, and understood clearly what was being said. In the twenty-first century even scientists tend to know Latin only by rote. For this reason, and because of the increasing participation of scientists using different alphabets, it is often suggested that the ponderous and often difficult binomials should be replaced with alpha-numerical codes that would be better understood by computers. It is hard to see this happening. 'We have close to a million species already assigned Latin names,' says Max, 'so any attempt to change the system would require absolutely vast retrospective work, taking lifetimes.' It is also alien to the human instinct for naming things. Not even a scientist would want to refer to number 53471c, rather than, say, *Brachytarsomys villosa*, or the hairy-tailed tree rat.

The number of taxonomic levels – main branches on the tree – has multiplied since Linnaeus, whose *genera* (plural of *genus*, the groups into which related species are placed) were extremely large. Again Max takes his example from insects. 'He had the genus *Papilio*, in which most of the butterflies were placed. Nowadays *Papilio* is only a group of swallowtail butterflies, and most of the others are in different genera because people have become aware of more species, and more differences, and as we've got more species we've had to come up with more categories in order to explain those differences. Since Linnaeus, people have been chopping the genera up smaller, and introducing other levels of classification.' At the end of my one-man seminar he lends me an undergraduate-level primer on the principles of taxonomy, which themselves seem to be evolving faster than any of the species under review. Armed with this, plus Joe's thesis and other learned volumes from various (mostly American) universities, I begin to construct some kind of basic understanding.

At the first level, life is divided into kingdoms. Linnaeus recognised just two – animals (*Regnum Animale*) and plants (*Regnum Vegetabile*). Two and a half centuries of exploration, analytical sophistication and genetic science have added four more – fungi, bacteria, archaea (single-celled micro-organisms) and protoctista (aquatic organisms including algae, seaweeds, protozoa and slime moulds). The kingdoms are then subdivided into *phyla* (plural of *phylum*), according to their basic characteristics. There are a great many of these, but the most important from the perspective of a mole-hunter are the *Chordata* – broadly, creatures with a spinal cord. Phyla are then split into classes. In animals there are six – mammals, birds, fish, reptiles, amphibians and arthropods (creatures with more than four jointed legs, including insects, spiders and crustaceans), each placed within the appropriate phylum. The classes then break down into *orders*. Typically there is no firm consensus about the exact number of mammalian orders, their names, and which species belong where, but the number seems to hover around twenty-six. They include, for example, the self-explanatory *Carnivora*, *Rodentia* and *Primates*, though of more particular interest to us is the order *Afrotheria*, which includes the golden moles.

Orders then separate into *genera*, or families, and the genera into species. It is these two last, genus – *Homo* – and species – *sapiens* – that comprise the scientific name. Humans therefore share with golden moles their kingdom (*Animalia*), their phylum (*Chordata*) and their class (*Mammalia*), but diverge at the level of *order*. As relationships go, it's pretty remote. Other relationships are becoming more or less distant as DNA testing and molecular biology are reshaping the tree. Branches are lopped

False gold – European moles trapped in an English garden. The pale one is the colour of marmalade, but no relation to the golden moles of sub-Saharan Africa

off, turned upside down and grafted back on different boughs. Golden moles, it turns out, are perfect examples of the unreliability of the old Linnaean method of grouping lookalikes. Their appearance, habitat and behaviour all so closely resemble the European mole's that their family relationship is obvious – obvious, that is, but wrong. In phylogenetic terms their shared common name could hardly be less appropriate.

The European mole, plague of field and garden, belongs to the order *Eulipotyphla*, which also contains the shrews (*Soricidae*) and hedgehogs (*Erinaceidae*) as well as forty-two species of 'true' mole (*Talpidae*). Golden moles are of the order *Afrotheria*, which contains several other species that bear misleadingly strong resemblances to the *Eulipotyphla*. Madagascan tenrecs, for example, could easily double for European hedgehogs. But the golden moles are acquiring relations as well as losing them. Elephant shrews, for example, were first described in the 1880s and given their common name because they looked so much like all the other known shrews – long noses, sensitive whiskers and an appetite for worms and insects. In the 1990s, however, it was revealed by genetic sequencing that they were not shrews at all but twigs on a faraway branch that contained not just golden moles, tenrecs and aardvarks but was a near neighbour of some of the most improbable relatives it is possible to imagine – manatees, dugongs, hyraxes, marsupials and elephants. It was not that European and golden moles had descended from a common ancestor, but rather that natural selection had adapted them to their similar environments, a perfect example of what scientists call *morphological convergence*. As Max says, physical similarity very often *is* a good indicator of family relationships (horse, donkey, zebra, for example, or the cat family), but you can see how unknowingly difficult life was for scientists like Linnaeus who had only the evidence of their eyes to guide them. A shrew is not always a shrew. A mole is not always a mole.

Calcochloris tytonis and its golden brethren constitute the family *Chrysochloridae*, and were originally described 250 years ago by Linnaeus himself. DNA and fossil evidence suggest that their clade, the order *Afrotheria*, first branched out some 100 million years ago. The world then was unlike anything humans have ever known. Antarctica seems to have been some kind of tropical paradise. Evolution had just produced the first bees, but ants still lay in the future, as did *Tyrannosaurus*, bats, butterflies and, a very long way down the track, humans. The earth itself was heaving with massive uncertainties. Shifting tectonic plates were tearing apart the great lumps of rock that would morph into South America and Africa, and it seems probable that the proto-Afrotherian, whatever it looked like, was isolated on the African side. A million centuries later, we can thank it for the aardvark, the elephant and the eponymous 'shrew'.

And also, of course, for golden moles. In the 250 years since their discovery, they have proved remarkably difficult to know. Gary Bronner attributes this to the remoteness and smallness of their ranges, and to the typical shyness of small blind creatures that spend their entire lives underground. This rules out casual sightings and makes them extremely difficult to track down. More pertinently perhaps (for we are talking about Africa), the abundance of more charismatic animals has denied them the attention they might have received in a poorer environment such as Britain's. (The truth of that is rubbed in, on this very day of writing, by a new book published in Britain by the Mammal Society on its top priorities for conservation – red squirrel, hare, harvest mouse, hedgehog, wildcat, pine marten and polecat.)

I am afraid that what I have written may sound more definitive than it really is. Taxonomy forever is in flux – written in sand, not carved in stone, and varying from source to source. What other scientific discipline has been at the same time so meticulous and yet so unreliable? For all their invisibility, golden moles are a good example, being passed like foster children from family to family. It was suggested in 1916 that they should be given an order of their own, but they were dumped instead in *Insectivora* – the order of insect-eaters. But diet is too loose a concept to be genealogically useful. Imagine lumping flesh-eaters together. Your Aunt Agatha would be in the same order as her cat. The whole order of insectivores, little more than a lumber-room for odd mammalian bits and bobs, was later cleared out and its denizens rehoused.

To begin with, the bracketing of golden moles with tenrecs rested on their unpronounceable teeth. When it comes to obfuscatory language, scientists need no lessons from lawyers. In the tiny mouths of golden mole and tenrec we find a real whopper – *zalambdodont*. It is an adjective, and it describes their molars. From the *Complete Oxford* – for it's beyond the scope of any single volume dictionary – I learn that 'zalambdodont' derives from the Greek words for the letter *lambda* and *tooth*, and that zalambdodont teeth have V-shaped ridges on them.

'Zalambdotonty', however, turned out to be one of those misleading physical coincidences like the elephant shrew's nose. As Gary Bronner notes, V-ridged teeth have occurred independently in several other mammals (solenodons, for example), so it is probably explained by morphological convergence rather than by common ancestry. In a further twist, this has turned out not to matter. I won't try to describe, or pretend to understand, how genetics confirmed what the teeth had first implied – that golden moles and tenrecs had so much shared history that they were a clade in their own right. But that is what happened. In 1999, therefore, from within the 'superorder'

Afrotheria, the new order *Afrosoricida* was born – the exclusive preserve of golden moles, tenrecs and otter shrews, now out on a twig all of their own.

Thanks to a paper by Gary Bronner, I am able to put a bit more meat on the bones. There are, he says, twenty-one known species of golden mole, all confined to sub-Saharan Africa. Despite their name (who could be surprised?), they are not all coloured gold. The family name, *Chrysochloridae*, derives from the Greek, 'green-gold', a reference to the 'iridescent sheen of coppery gold, green, purple or bronze' on their fur. This chimes very nicely with the 'metallic reflections' noted by the *British Cyclopaedia of Natural History* in 1836. The Victorian authors, however, had complained that a stuffed skin could give them 'no idea of what the living animal is like'. Here Bronner is more helpful. Despite variations in size and colour, they all look very similar. He sends me back to the dictionary to find out what 'fusiform' means (lozenge-shaped body tapered at both ends), but otherwise his word-picture has a Dürer-like precision. The forelegs are short and powerful with 'pick-like' claws. There are no external ears, eyes or tail. On the densely furred pelt, the woolly under-fur is protected by a moisture-repellent overlay of 'guard hairs'. Beneath all this lies a thick tough skin which is particularly robust on the head, and the muzzle has a leathery nosepad to protect the nostrils. Underground the animals tunnel like machines, with upthrusts of their flattened heads and downthrusts of the claws, leaving a ridged 'wake' on the surface as they go. A few of them also throw up molehills, thus adding to the confusion with the European *Talpidae*.

When it comes to the exact object of my quest, however, I am hardly better off than the learned authors of the *British Cyclopaedia*. At least they seem to have had a skin to refer to. For the umpteenth time I turn to Alberto Simonetta's paper of

1968. One of the most interesting things at the time seemed to be that the discovery, in the author's own words, 'extends by over 750 miles to the East and considerably to the North of the known range of the [golden mole] Family'. Thus it would seem to be either a known species that had gone walkabout or a new one on its own territory.

If it was a known species, then analysis of the ear-bones would swiftly reduce the list of suspects. It hinged on the *malleus*, the small hammer-shaped bone that transmits vibrations from the eardrum to the middle ear. In some golden moles, I learn, this has a 'hypertrophied ball-shaped head'. Back I go to the *Complete Oxford*. *Hypertrophy*, I discover, is the increased volume of body tissue resulting from an enlargement of the cells (as opposed to *hyperplasia*, which is an increased *number* of cells). In some other golden moles, too, the head of the malleus is 'elongated and club-shaped'. As the pellet specimen had neither of these characteristics, it eliminated every species bar those of the genus *Amblysomus*. After that, it all came down to the teeth. I am really struggling now. Most of the *Amblysomus* species apparently have what the professor describes as 'a more or less well developed talonid' on their molars (dictionary again: a *talonid* is a flattened cusp). The specimen has no talonid at all, thus reducing the possibilities to three. After that, I confess, I am pretty much lost. My comprehension scrabbles like fingernails on rock, then slides gracelessly into the abyss. It is fashionable at the moment to talk about 'journeys'. They are always being embarked upon in TV cookery contests or reality shows, and imply some kind of glorious ascension from darkness into light. My stubbornly unscientific brain takes me in the opposite direction. I *think* I understand something, then I lose my bearings in a fog of detail. So it is with the teeth. It has been a long journey indeed, from the worldwide sweep of *Animalia* right down to the microscopic detail of a mole's dentition. Simonetta gives us every conceivable datum – 'length from tip of lower jaw, teeth excluded, to occipital cordyle', 'breadth of ascending process', 'maximum breadth at tip of angular process', 'length of dental row at alveolar margin', and so on. The dictionary too now rolls over and waves its legs in the air. Simonetta goes on, with ever finer detail, for twenty-eight pages, only a fraction of which I am able to follow. The shape and length of the jaw are somehow different from all previously known species – at the moment that's all I can say. But it is enough for Alberto Simonetta to conclude that the specimen is unique. It is customary in taxonomy to name a species after its discoverer, or after someone the discoverer would like to honour. But here the commemorated hero was to be neither Simonetta himself nor any of his esteemed friends and colleagues. The honour instead would belong to the beneficent deliverer of evidence, the consumer and regurgitator of the new mole's last remains. From the barn owl, *Titus alba*, sprang forth its own dedicated species, *Calcochloris tytonis*.

Thus do I manage to achieve some rudimentary understanding of nomenclature and taxonomy, and of the elephant traps that await anyone with ambitions to demystify the processes of nature. Several of the species mentioned in Simonetta's paper, including *tytonis* itself, have been shifted from one genus to another since its publication in 1968. Right across the tree, branches have been swishing in the storm of new evidence as species have found new relatives among the living and the fossilised dead. It's not something that we read or hear much about – an arcane process that continues unseen in the back offices of natural history museums and makes headlines only in learned journals. I love it, though. I love the suppleness of science, its willingness to change its mind and head off in new directions. That surely is the best and most powerful validation of objective study, its great and decisive advantage over dogma. It is why Copernicus triumphed over the Church, Darwin over Wilberforce, Huxley over Owen. The more you look, the more you realise how much there is to see. A golden mole opens its mouth, and therein lie all the miracles and mysteries of creation.

The miracles and mysteries of human civilisation, however, remain to be understood. At the end of a rain-sodden week I make another trip to London. Most of the things that can go wrong do, though I am spared a suicide. The railway line crosses the Cambridgeshire Fens, a dead-flat landscape of reclaimed seabed, scored with dykes, that a very few people love and many more find melancholic. For some, the temptation to step in front of a London-bound train is irresistible, and it is a tragically frequent cause of delay. Deliberate self-harm is another peculiarity of my own species that finds no echo in nature, though the myth of mass suicide by lemmings still persists (in reality, their deaths by drowning are the accidental results of over-ambitious sea crossings). My day requires an early start and a missed breakfast. The normally sedate forty-five-minute drive to the station is turned into a mad, heart-pumping dash by a long tailback from roadworks where 200 metres of carriageway have been coned off for the convenience of one man and a shovel. The price of my train ticket is all that you would expect from Europe's most expensive rail network, though at least I get a seat (not a privilege available to those who board at intermediate stations). Insofar as it concerns itself with news, the paper I buy is full of grim stuff about economic crises and the moral elasticity of bankers force-feeding themselves with other people's money. London itself does nothing to lighten my mood. If the managers of the underground system mistreated cattle in the same way as they mistreat commuters, then they would spend the rest of their lives in jail. Despite all this I arrive in good time for my meeting, but the person I am seeing is half an hour late. This delays my homeward journey until the evening rush hour, but I get to the station in good time to claim a seat on the train. Then I glance up at the platform indicator. The service has been cancelled; there has been a suicide; passengers are advised to find a different route. This means another nose-to-armpit underground crush to a different London station, then a long, roundabout crawl through rural halts that adds another two hours to my journey. For an hour and-a-half of this, I have to stand. When finally I get home, I feel a desperate need for an alcoholic drink.

But there is something else. The red light on the telephone is blinking. A message. Wearily I tap the Play button, and there it is. A bit of fuzz and crackle; the bathroom echo of a bad line over distance. And then the voice. It is a little faint but the words, spoken with a soft Italian lilt, are as mellifluous as birdsong.

'Hello,' it says. 'This is Professor Simonetta.'

CHAPTER ELEVEN

Valete Et Salvete

Next morning I call the professor back. If he is surprised by my reverential tone – I realise I am actually *bowing* over the telephone – he does not reveal it. But there is a twinkle in his voice that suggests humour. He is in his eighties now but apparently still busy. Better still, he speaks good English. He listens while I pour out my story, which I tell in a confused rush with little sense of order or economy. How can I convince him that I am basing a whole book around his tiny fragment of golden mole? It must surprise him that a layman had even heard of it, let alone turned it into a quest. He chuckles. There is much he could tell me, he says. Many stories. But does he have the Somali golden mole, *Calcochloris tytonis*, the world's rarest known mammal? Ah! Well, of course he doesn't have it *personally* . . . But why don't I come to see him in Florence? Then he can tell me. There are so many stories . . .

It is 9 a.m. on 22 June 2012, a date inked into my diary. We agree to meet in the third week of September. The professor will send me his address. I thank him with near-idiotic profuseness and put some champagne on ice. Florence! Venice alone might beat it as my favourite city, but it would be by only the smallest of margins. If I don't find the mole, then at least there will be Leonardo, Michelangelo, Boticelli and rare Florentine steak. As it happens, blood, meat and mortality have been weighing heavily on my mind. An old friend has recently died and I have just returned from visiting a slaughterhouse. The two strands of thought have tangled themselves with the mole into an unpickable knot.

We are not clever about death. In the developed world we are living longer than at any time in our species' history. But the very remoteness of death, and the tidy packaging of it by morticians, means that – also unlike any people in history – we keep it out of sight. Many people will die without ever having seen a dead specimen of their own kind, or even having applied the word 'death' to another human. We speak rather of 'passing away'. The sensitivity extends to pets, which may be posthumously anthropomorphised at special crematoria or cemeteries. But many other animals, of equal or greater intelligence, will have no memorial beyond the thickening arteries of those who have cooked and eaten them. At the slaughterhouse I have spent a morning watching lambs die. They are stunned by electrodes placed across their heads, then have their throats cut. A mechanised conveyor delivers them in a nose-to-tail stream to the slaughterman, who kills them at the rate of one every nine seconds. I follow the whole process from the delivery of live lambs at the lairage gate to the dispatch of dressed carcasses from the refrigerator. Placing logic before emotion, I remind myself that squeamishness is not morality. I belong to a meat-eating society whose gods granted it dominion over all other species – gods, indeed, to whom animals were ritually sacrificed. We kill and we eat. It's been that way for millions of years, though for all but the last small fraction of that time people were far closer to the bloody realities of what they ate. The act of slaughter was visible and common to every community, not hidden away like it is now. In no century earlier than the twentieth would a writer have experienced as much difficulty as I did in persuading a slaughterhouse to let him through the door.

It made me think about the value judgements implicit in the theme of this book. If someone were to find and kill a living example of *Calcochloris tytonis*, my anger would burn holes in the page. And yet I accept the deaths of lambs without qualm. What, then, is my moral position? Do I have one? Another awkward question: is it worse to kill a rare animal than it is to kill a common one? Moral philosophers – those who talk about animal rights – would say no. A life is a life, and it is not for us to judge the value of one over another. Indeed, on the utilitarian principle of 'the greatest happiness of the greatest number', they might argue that numerical supremacy and moral value marched in lockstep – a dangerous principle in the wrong hands. As I said in Chapter Nine, 'speciesism' – the notion that individuals should be favoured or disfavoured purely on the basis of their position on the phylogenetic tree – is abhorrent to them. This plainly rules out any idea of animals as property, which means goodbye to pets as well as to meat on the plate. It means also a moral dilemma wherever the interests of people are in opposition to other species'. Pre-linguistic infants, senile and insensible humans are routinely invoked as proof of our inconsistency. If animals have similar or superior abilities to these, then why do they not enjoy the same moral rights?

It is a neat but ultimately unsaleable argument, best left to the debating society. Perhaps I should be capable of a more enlightened outlook, but I am not. As a leather-shod meat-eater kept clean and healthy by products tested on animals, I have no moral high horse to ride upon. But I don't believe this should commit me to the outer darkness. Moral precepts are not immune to circumstance. Like everything else in a changing world, they answer to market forces. In the middle of the nineteenth century, when the supply of wildlife seemed inexhaustible, it might have been no more unreasonable for Roualeyn Gordon-Cumming to roof his bivouack with an elephant's ear than for me to grill a cutlet. Adventurers then had a concept of novelty, but no appreciation of rarity or endangerment. How could they? For us, who must wrestle with the consequences of all that humankind has done, these are wholly new ingredients in the moral mix. So I repeat the question: is it worse to kill a rare animal than a common one? Answer: yes. It might not please the anti-speciesists, but most people of uncomplicated view would think it worse to extinguish a species than to snuff out an individual. Like it or not, this imposes a progressive scale of values. Bio-ethics is not immune to the laws of supply and demand. I frankly admit that I would have no interest in Professor Simonetta's owl pellet if its contents were commonplace.

Saving the savable is the very lodestone of wildlife conservation. What to preserve, how and where to preserve it, are questions that are easier to ask than to answer. It is easy to say what is desirable; much harder to know what is possible. The number of variables – population, fertility, habitat, food supply, climate – make conservation an issue of high risk and complex calculation. It is here that phylogenetic trees might be useful. Like climate models they can tell us something about where we are going as well as where we have been. I have at my elbow a grey slab of academic text bearing the names of eleven distinguished authors from scientific institutions in three continents. 'Phylogenetic trees and the future of mammalian biodiversity' has lain on my desk for a month or more, and I can put it off no longer. Today is a rare phenomenon, a day of searing heat in the weird British summer of 2012, when northern Europe played chicken with the jet stream. Out I go into the sunshine to park myself under a maple tree with a notebook, a pink highlighter and a jug of iced elderflower cordial. Outdoor reading has its rituals and its distractions. On a morning such as this, even a largely denatured England seems frighteningly alive. It is one of those days that recalibrates the eye and makes you look afresh at the over-familiar view. There are shades of green here that would have confounded Cézanne. The grass is speckled with daisies and clover; a spider's thread catches the sun and becomes a tiny laser beam; a brown butterfly tumbles past, as if churned by some hidden current in the still air; a fledgling thrush hops guilelessly into view. There are ants, a bumblebee, the inelegant flap of a woodpigeon moving from beech to ash, the rhythmic *chomp* of a baler in a hayfield. A ladybird settles on my arm. Banal thoughts sometimes are difficult to resist. I stroll down to the dyke outside my garden, a narrow trickle that feeds a stream that joins a river that flows down to the sea, which, by way of gulf, delta and estuary, joins me to every place on earth. Thus do I now see the brown butterfly, the bumblebee, the ant, the pigeon, the tractor driver, me – all of us rafting on the same river of life, eddying and moving on.

I abandon the aquatic metaphor and return to the branching structures of the phylogenetic tree. By noticing where branches end, or where they divide into twigs, scientists can work out which kinds of species in which kinds of environment are more or less likely to thrive or to diversify. Some of their conclusions would have been no surprise to the hunter-slaughterers of the nineteenth century or to the penitent butchers of the early twentieth. The tropics both literally and metaphorically are the hothouse of mammalian diversity, and – *pace* Ol Pejeta – concentrations are especially high in Africa's Great Rift Valley, where they peak at more than 250 species per 100-kilometre square. There are peaks also in Amazonia and, say the authors of 'Phylogenetic trees', 'in an arc running from the Himalayas into south-eastern Asia'. This brings economics into the argument. Is it better to concentrate on areas with a relative abundance of life, relatively easy to protect; or to firefight where life is on the brink?

While a mob of unidentifiable black insects buzzes around my head, I read about 'sister clades' – groups of closely related species within a genus. They are most common in genera with high populations and large litters, which may explain why 920 species have 'mouse' in their name, and only three 'giraffe' (the camelopard itself, plus a seahorse and a catfish). Having few close relatives is a strong indicator of risk. Less surprising is the news that wild mammals tend not to thrive in proximity to humans. The author of Genesis might have been a great stylist but he was no great seer. Dominion we have had for millennia, but we've yet to acquire the knack of looking forward with our eyes open. No biblical prophet ever came down from the mountain with a premonition of global warming, or even with any idea that there was a globe to be warmed. Still less did they have any idea of the destructive past. Some 10,000 years before the time of Christ, at the tail-end of the Pleistocene, climate change had already caused a mass wipeout. The species that vanished were mostly large – cave-bear, mammoth, mastodon, smilodon (sabre-toothed cat), woolly rhinoceros, giant beaver, giant sloth and many more, all now represented by stump-ends on the tree. The current mass extinction threatens to be similar. On average, say the authors of 'Phylogenetic trees', declining mammals are an order of magnitude heavier than non-threatened ones. Thus the wood mice, which helped fill the void after the disappearance of mammoths, remain secure in their many niches while rhinos have struggled to cling on. The special vulnerability of big animals is pretty obvious. The smallest creature I have ever heard of being shot at was a spider, at which my grandfather (who enjoyed his beer) aimed an airgun so unsteadily that he shot himself in the hand. Hunters on the whole prefer something easier to aim at, so it is the big beasts that have attracted the spears and bullets.

A big animal is a more serious loss to its species than a small one. Megafauna have smaller litters of larger (and so more vulnerable) offspring, which take longer than small animals to mature. A female black rhinoceros, for example, will reach sexual maturity at between four and seven years; a male at between seven and ten. Calves are born singly, after a gestation of fifteen to sixteen months, and then take two years to wean. The birth interval is between two and a half and four years. Now compare a wood mouse. Sexual maturity at two months; gestation twenty-one to twenty-six days; up to four litters a year, four to seven babies each time. Bigger animals also tend to be more specialised in their territory, and to need more of it for each individual, which makes them especially vulnerable to habitat loss as well as to hunting. Specialised habitats also tend to be localised, which puts whole populations at risk. This is why rhinos and elephants can go missing from entire countries.

There is actually a measurable threshold – 3 kilograms – at which the size factor kicks in. This is about the weight of a smallish red fox, the smallest animal regularly hunted for sport. But you only have to look at the *Red List* to see that smallness is not an absolute defence – at best a tin hat rather than a bomb shelter. Innumerable diminutive species driven from their ranges, out-competed by introduced species or robbed of their habitats, are heading for the cliff edge. Because it is rather easier to prove the non-existence of a large animal than a small one, much of the search effort for highly endangered or possibly extinct species is for animals of less than 3 kilograms. But it's the big stuff – the wild ox and horse, the thylacine – that gets the most attention.

Near the temple of Wat Phnom in the Cambodian capital Phnom Penh stands what many visitors might suppose is a larger-than-life statue of a charging bull. With its huge curved horns and corded musculature, it looks far too big for its tiny railed enclosure, like a shark in a shrimp net. Tourists kneel to get the most dramatic, head-on image of this seemingly mythical monster. Braver souls have taken almost absurd risks to track down the real thing. Giant it may be, but mythical it is not. The kouprey, or wild grey ox, *Bos sauveli*, is the biggest and most virile of Cambodia's several national animals. Its short acquaintance with humanity contains all the elements of tragedy and near farcical comedy that can descend on science like moths on a wardrobe. There are times when you can't see the substance for the holes. Does the kouprey still exist? If it does, is it a bona fide species of genuine scientific value, or some kind of wild/domestic cross-breed of no more than passing interest? *Passing*, certainly, is the *mot juste*.

The kouprey is, or was, every bit as imposing as the statue implies. It stands over 6 feet tall at the shoulder and weighs up to 2,000 pounds, or 0.89 imperial tons. It was 'discovered' – i.e. brought to the attention of western zoologists – only in 1937, and was believed to range through northern Cambodia and parts of Laos, Vietnam and Thailand. An American zoologist, Charles Wharton, managed to film some in 1951 but that, pretty much, was that. The last verified sighting was in the 1960s. Wharton did catch five live specimens in 1964 but the mission ended in disaster when three of them escaped and the other two died.

Cambodia for most of the time since has not been a happy hunting ground, or indeed any kind of hunting ground, for zoologists. The most heroically mad expedition was mounted in 1993 by the daredevil American journalist Nate Thayer, whose apparent blindness to danger earned him comparison with

Colonel Walter E. Kurtz, Marlon Brando's crazed maverick in *Apocalypse Now*. According to an account by two of the participants, Robert K. Brown and Robert MacKenzie in *Soldier of Fortune* magazine, Thayer led a terrifying twenty-five-strong posse including journalists, jungle trackers, former Khmer Rouge guerrillas, a British photographer, a Thai television cameraman and an Italian expert on camels. Not all the journalists might have passed drug tests, but they did have the foresight to complement their notebooks with walkie-talkies, AK-47s and a rocket-launcher. Together, riding on elephants and eating lizards, they set off into one of the most inaccessible parts of one of the most dangerous countries on earth, successfully tempting exhaustion, illness and the curiosity of the Khmer Rouge. They survived, but not a kouprey did they see. And not a kouprey has anyone seen since.

Worse was to follow. In 2006, scientists from Northwestern University at Evanston, Illinois, citing DNA evidence from two kouprey skulls, argued that *Bos sauveli* was not a proper species at all but the bastard offspring of domesticated banteng (another species of wild ox) and zebu cattle (otherwise known as Brahmins, progenitors of Ol Pejeta's Borans). 'It is surely desirable,' said their team leader, 'not to waste time and money trying to locate and preserve a domestic breed gone wild. The limited funds available should be used to protect wild species.' This stuck like a thistle in the craw of conservationists who had been trying for decades to find the kouprey, and who now found themselves relegated from legitimate scientific investigators to a mere breed society. This is the great dilemma of the Anthropocene. When man and nature dance together, it is not always clear who leads.

Happily in this case the tune would soon change again. By a miracle of synchronicity, a fossilised kouprey skull suddenly turned up. As is the way with fossils, its age was a bit slippery, probably somewhere between the late Pleistocene and early Holocene, giving it a range of between 5,000 and 125,000 years. But it didn't matter. As the Northwestern team-leader himself conceded: 'You can't have a fossil kouprey skull if the kouprey is a recent hybrid.' Scientists from the National Museum of Natural History in Paris then did some further analysis and concluded that a female kouprey and a male banteng had mated some time during the Pleistocene, and that it was this – not modern hybridisation – that explained the discovery of kouprey DNA in bantengs. The dancers reversed, *Bos sauveli* got its status back and a branch high up on the phylogenetic tree stopped quaking.

But the consolation was academic. The kouprey recovered its dignity but not its life. Loss of its forest habitat had taken the inevitable toll. So had diseases transmitted from domestic cattle.

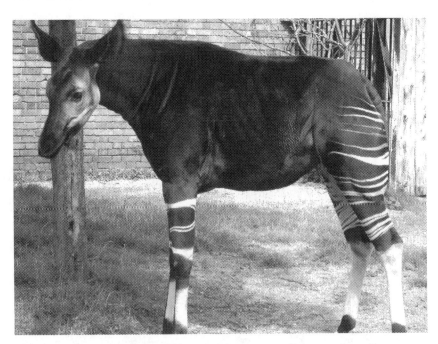

Okapi at London Zoo. The species was not discovered until 1901, when first reports of its existence were dismissed as a hoax

But the decisive genetic cleanser was hunting – initially for meat and then, as scarcity ramped up the value, for skulls and horns. The IUCN *Red List* reports horns for sale at Ban Mai, on the Thai–Lao border, for up to 12,000 US dollars a pair. It is hanging fire on declaring the species extinct, but with no confirmed sightings for nearly fifty years the time cannot be far off. 'Its extinction, if not yet upon us, is certainly sealed,' it says.

The death of a newly discovered species is practically a self-fulfilling prophesy. My mind goes back to the Natural History Museum and the 'new' Madagascan carnivore, Durrell's vont-sira – largely forgotten since its Warholian fifteen minutes of fame in 2010. When a species escapes detection for as long as this one did, the likeliest explanation is that it is vanishingly rare. So it was with the kouprey. So it is with its near neighbour the saola. Earlier I described it as an antelope, which is what it looks like, but *Pseudoryx nghetinhensis* is actually closer to cattle. One biologist described it as 'a cow that behaves like a goat'. When it was discovered in 1992 it was the first previously unknown large mammal to have been found anywhere in the world since the kouprey in 1937 (which in its turn was the first since the okapi in 1901). Its fate is testimony to both the strength and the weakness of human interference. Such is the dominance of *Homo sapiens* over his environment that the power of gods looks feeble by comparison. Mythology cringes in the shadow of the supreme creator and destroyer. There is nothing ironic in the title of Mark Lynas's challenging but well-argued book, *The God Species*. He intended it to be taken literally. In the orthodoxy of the anti-GM, anti-nuclear green mainstream, 'playing god' is the ultimate act of hubris by which we will render ourselves into dust. Like other independent thinkers, Lynas, once a powerful voice for the green consensus, now argues the opposite. Playing god, in the sense of being intelligent designers, is essential if creation is not to be irreparably harmed. The earth is out of kilter, but by calculated and cooperative acts of benign intervention we can help it to regain its stability. Essentially, however, we're talking about damage limitation. Human genius might find some technological fix to the life-threatening problems of climate change and ocean acidification, but – aside from the wilder fringes of scientific fantasy – the evolutionary clock cannot be turned back. We have several very effective ways of exterminating species, but no very effective way of bringing them back. Unlike the kouprey, the saola is still clinging on, but its grip is weak. Numbers already may be so low, says the *Red List*, that no viable populations remain. Following the kouprey through the limelight and out again, the saola is into the last act before the final curtain.

It is a classic victim of sizeism. Throughout its range in the Annamite Mountains, all species heavier than 20 kilograms have been hard hit. Even animals as commonplace as muntjac, Sambar deer and wild pig are rare here, and wild cattle, elephant and tiger are all but extinct. One way or another it is hunting that has done the damage. Much of it has been straightforwardly for meat. The IUCN reckons that eight million Vietnamese people 'with the propensity to eat wildlife' live within 100 kilometres of the saola's forests. But this is not the only, or even the worst, threat to its existence. The supply line for many of the animal ingredients in Chinese traditional medicines begins here, in these very same forests. Although the saola has no direct role in the Asian or Chinese pharmacopeia, it makes very little difference. The snares are indiscriminate. The cost in lives can only be guessed at, but the IUCN has come up with some figures that hint at their enormity. More than a billion people live in China, and more than seventy million in Vietnam, a vast, proximate and hungry market for animal extracts and body parts, with increasing wealth and a growing population only intensifying the demand. (It is a myth that wildlife trafficking is linked to poverty.) The IUCN concludes that every square kilometre of the saola's range has snares in it, and that hunting in some areas is so intense that it amounts to 'many thousands of snare-nights per square kilometre per year'. As stocks of deer, civet and pig far exceed those of saola, hunting will continue until well after the saola is gone. The process is accelerated by logging, farming, road-building and hydropower, all of which are taking large bites out of the forest. In the circumstances, we should perhaps be surprised that any saolas survive at all. Not even captive breeding can stop the rot. Twenty or so have been caught alive, but eighteen of them died and the other two had to be released. Meanwhile, scarcity has only increased the saola's value as a trophy in Hanoi as long ago as 2000, horns were being quoted at 600 US dollars a pair. When extinction comes, as it must, there will be a few headlines around the world, but the obituaries for a species few people have heard of are likely to be brief. Like the kouprey, the saola will simply slide off the page and be forgotten.

The word 'depauperate' is often used to describe ecosystems that have lost important species. If it applies to Vietnam, then it applies just as much to Britain. This is a source of discomfiture. Environmental politics is bedevilled by the absence of good international role models. The charges of hypocrisy flung at rich countries by the poor are easy to make and hard to refute. Latecomers to the feast are not easily charmed by western nations patting their stomachs and telling others it's their duty to stay thin. As it is with climate change – do as we say, not as we have done – so it is with wildlife. Having wiped out most of our own, we have found a sudden interest in dissuading others from doing the same. The cause is good, but the glass house is a tricky place from which to launch it.

The last person I heard complain of 'depauperatism' was a
zoologist trying to make a case for the reintroduction of wolves
in Scotland. With Ol Pejeta's lions and livestock in mind, this
might not be as crazy as it sounds. Or not quite. In England
the government's official conservation adviser, Natural England,
declares its intention 'to conserve and enhance the natural environment, for its intrinsic value, the well-being and enjoyment
of people and the economic prosperity that it brings'. The
language is bland but the idea is radical. That bracketing of
natural and human interests would have been anathema to
Natural England's predecessor, the Nature Conservancy Council.
For them, 'people' were pests to be excluded wherever possible
from 'the wild'. The old way was a powerful force in the polar-
isation of interests – wildlife versus crops – that characterised
the post-war agricultural revolution and guaranteed that there
would be more losers than winners. Who would have imagined
that the finger-in-the-dyke UK Biodiversity Action Plan would
need to include the likes of hedgehog, hare, harvest mouse,
skylark, cuckoo and toad? It was unimaginable, but it happened
– another fine example for others not to follow. For decades
after the Second World War, the countryside suffered under a
kind of apartheid – one place for wildlife, another place for
man. Displaced by their human oppressors, animals were
confined to ghettos where they lost touch with each other.
Biodiversity became bioconformity, and nature sank to its lowest
ebb since the Ice Age. It was a kind of taxonomic multicultur-
alism in which a home for one population too often meant the
displacement of another, and in which wildlife conservation
turned into ambulance-chasing. How can we save the red
squirrel? The dormouse? The stone curlew? This was nostalgia
– a misty-eyed yearning for the good old days of Ratty and
Mole, and for the countryside of Clare and Constable – and

nostalgia is no substitute for the kind of clear-headed thinking that will actually put fur on bones. This may look parochial – an English writer bemoaning the state of an industrialised landscape in which a corncrake is about as welcome as a rat in a Pot Noodle factory – but it's not. The problems are the same wherever in the world man and wildlife clash. You can't conserve what is not conservable. It was French medics in the First World War who formalised the system of *triage*, by which they sorted patients by order of need. As it was for men in their field of war, so it is for animals in theirs. There are species that can look after themselves, species that can be saved if they receive the right attention, and species that are beyond hope. On this basis it could – some would say *should* – be goodbye to the giant panda and goodbye to the red squirrel.

England's long campaign to preserve the red squirrel from the invading grey has been an exercise in futility. The grey – which, unlike the red, is happy to cross open ground – is better adapted to the altered countryside. It is also bigger, fiercer and wholly illiberal in its attitude to rivals, and carries the squirrel pox virus that kills the red. The Prince of Wales, uncrowned king of the dreamers, wanted the red squirrel to be adopted as the nation's mascot, and declared that it was 'absolutely crucial to eliminate the greys'. I once had a conversation about this with the BBC wildlife presenter Chris Packham, who simply pointed through the window to the municipal park opposite, where grey squirrels were swarming through the trees. The questions hardly needed asking. How could you expect to wipe out an entire species? Who would pay for it? How could it be imagined that rescuing red squirrels, which need continuous tree canopy, was simply a question of slaughtering greys?

'The bigger issue,' said Packham, 'is that our countryside is in ruins, and our habitats are in catastrophic decline. You can moon and coo all you like over nursery-book favourites like dormouse and red squirrel, but it won't bring the countryside back. What it actually does is divert thin resources away from where they would do most good.' It is a stark truth. Every animal needs its natural habitat. If we can't provide it, then the species is condemned. This is why animals with narrow ecological tolerances, like red squirrel and giant panda, are at higher risk than generalists of more catholic taste. Historically in Britain, 'habitat' has meant a network of small and isolated nature reserves, ecologically fragile and in the long term unsustainable. Even in countryside as hostile as this, you have to think in terms of the landscape as a whole, not just isolated survival bunkers. It is all very well for farmers to join stewardship schemes that pay them for hosting wildlife, but the good intentions remain unfulfilled unless they can form an unbroken chain of neighbouring farms that allows movement across the land. This is a basic principle of the Ol Pejeta model, which encourages migration in and out of the conservancy. The scale may be bigger, but the thinking is the same – a linking of biological storehouses that opens nature's arteries and circulates the blood, releasing the patients from intensive care. It's not only animals that have to adapt to change. Wildlife managers sometimes need to revise their thinking too.

The reintroduction of locally extinct species into their original or similar habitats is the alpha and omega of rescue conservation. Again Ol Pejeta leads by example with the black rhinoceros. The reintroduction of scimitar-horned oryx, previously extinct in the wild, into the Dghoumes National Park in Tunisia is another huge international triumph, deservedly accompanied by the sound of trumpets, a perfect demonstration of the value of captive breeding. Good zoos now do not take

from nature. They give. There have been many other examples
– black-footed ferret in the US, Mexico and Canada; beaver
in Sweden; Zanzibar red colobus monkey on Pemba Island,
Tanzania; mountain gazelle and Arabian sand gazelle in Saudi
Arabia; Arabian oryx in Oman; Przewalski's horse in Mongolia.
They vary in their provenance. Ferret, gazelles and horse were
all captive-bred. Monkey and beaver were translocated from
the wild. They all breathe gently on the guttering flame –
pinpricks of light maybe, but in gathering dark a pinprick
shines like a beacon. As I write, a trial reintroduction is under
way of beavers in Scotland, the only part of Britain that has
much space unoccupied by humans. For years, zoological fundamentalists have had their eyes on it as the perfect place to
reintroduce the European lynx and wolf. This is a good illus-
tration of the kind of mess we get ourselves into when we
destabilise the fauna. Zoologists remind us that humans in
Stone Age Britain were outnumbered five to one by bears, which
lingered until some time between the eighth and tenth centuries. Lynx were thought to have become extinct at the
beginning of the fourth century, but bone analysis now suggests
that they survived in Scotland for another thousand years. The
last wolf probably died in around 1700, though there are plenty
of legends that place the event earlier or later (according to the
most persistent of these, it was killed near Inverness in 1743,
allegedly with two local children in its stomach). Without top
predators the fauna is incomplete, which is why fundamentalists want to bring them back. But what hope is there? Even the
beaver trial took ten years of political campaigning to achieve,
and received a sniffy response from farmers who dismissed it
as a 'costly luxury'.

   If the vegetarian beaver has had to wait so long for admission,
then what chance is there for bear, lynx and wolf? For bear we

can confidently answer: none. For lynx we can predict a fair chance, and for wolf at least something on the upside of nil. The case in their favour is that wolf and lynx both do what big carnivores are designed for – they eat herbivores. In Scotland this means deer, which, in the absence of predators, have become a pestilence. In England, too, they are a scourge of crops, woodlands, gardens and cars. No accurate figures are kept, but the best estimate is that deer every year cause 34,000 road accidents in Scotland and 8,000 in England, injuring more than 400 people and killing ten. Despite heavy culling, the combined population of the six resident species – red, roe, fallow, sika, muntjac and Chinese water deer – is heading towards two million, and likely to increase by 10 per cent a year.

But what are we to do about it? Lynx would certainly eat a few, but a few is all it would be. Computer modelling suggests Scotland could support a population of up to 450, which would be enough to boost eco-tourism but not enough to stem the tide of deer. If the lynx is to be reintroduced, then it will have to be for its own sake, not as an organic pest controller. There is a moral and ethical point here. Reintroductions of locally extinct species are allowed under European law only if the original cause of their extinction has been removed. In the lynx's case this means persecution by humans, which is precisely why zoologists say we have a duty to bring it back, the only meaningful apology for past misdeeds. There is a political consideration too. Unless we restore our own landscape and the animals that belong in it, then where is our authority to lecture Laos and Vietnam on theirs?

Wolf-talk has been going on for decades – zoologists arguing in favour, and Highland sheep-farmers, with common sense on their side, questioning the lupophiles' sanity. For years the political wind blew the wolf's house down. It was not even

believed that it could have a significant impact on the deer. But
the ground shifted when evidence from America's Yellowstone
and other national parks suggested that the predictions had been
overly pessimistic. It might take a while, maybe sixty years, but
wolves and deer would settle into a balanced predator–prey
relationship in which deer would be reduced by more than 50
per cent.

Scientists also tested public opinion. Urban people, who
mistakenly thought that the principal risk from wolves was to
humans, were slightly more in favour than country folk, who
correctly understood that the risk was to sheep, but there was
an overall majority in favour of the wolf. The peculiar economics
of their industry means that even farmers are not as hostile as
they used to be. Without subsidies they would make a loss on
every animal sent to market. In the past, when subsidies
depended on the number of animals they kept, a lost sheep
would mean a direct hit on the farmer's pocket. But subsidies
now are paid simply for grazing the land, regardless of flock
size, so a dead sheep is a shame rather than a catastrophe. If
you factor in compensation for losses to wolves and profit from
eco-tourism, then the issue takes on a different shape.
Nevertheless, opinion remains polarised. Some scientists see
the wolf as an obligation; others as a step too far. Whichever
way the pendulum swings, there is no possibility of its swinging
very fast. South of the border, where much of the country gets
no closer to a wilderness than the grass on a roundabout, the
outlook is different. You would more easily reintroduce the
pillory than big carnivores to Middle England. If deer are to be
kept down, and if wildlife agencies are to make good their pledge
to reverse the degradation of broadleaf woodlands, then they
will have to rely on fence and bullet. Another man-made
problem, another human tweak.

On a world scale, the Highland wolf hardly registers as an issue. It's not a rhinoceros, or a snow leopard or a red panda. It's not threatened by global extinction – for the IUCN it is a species of least concern – and not many people would include it on their list of furry favourites. In fairytale, fable and folklore, only the rat wears a blacker hat than the wolf does. But it's a useful example because it shows how complex is the reckoning that must be done before an absent species can be put back where humans think it belongs. And it's a useful provoker of thought. As a top predator capable of maintaining equilibrium between the hunter and the hunted, its ecological function is easy to understand. But there are many other species on the IUCN *Red List* – threatened, endangered or critically endangered – so obscure, so small or so weird that there is no public awareness or concern for their fate. There will be more about these later. For now, they bind me to my purpose. Apart from myself, Gary Bronner and the man who discovered it, who in the world gives a zalambdodontic molar for the Somali golden mole? Who in the world knows anywhere near enough about the species, the habitats, the ecosystems they think should be saved? Think of the pine marten, another animal persecuted to near-extinction in Britain. It is now seen so rarely that, according to the People's Trust for Endangered Species, many people think it is a bird.

I have seen *la martre* in France but never in Britain. A few days ago I did have an encounter with its commoner cousin, the stoat. I looked out of the window and there it was, the ultimate exemplar of sleekness and glossiness, up on its points and curving its trunk with all the grace of a dancer in the limelight. There was no prey for it to chase, and nothing to threaten it, but still the dance went on. A behaviourist might have discerned some purpose in what it was doing, but I could see nothing

beyond the sheer joy of uninhibited movement, a joy atavistic-
ally shared by an observer made suddenly aware of the pleasures
of summer. I stood for a while after it had gone, waiting for it
to return but knowing that it wouldn't.

CHAPTER TWELVE

Unbelievabilia

I have a friend whose emails are always headed just 'Stuff'. By 'stuff' he means items of improbable interest, informational scraps that have caught his attention and come his way by chance. It is entirely random. Stuff can be jokes, quotations, absurdities from newspapers, news of acquaintances divorced, dying or falling off ladders. Mostly, though, it is *unbelievabilia*, odd bits of informational detritus scooped from the daily swill of news, stuff that it takes a beachcomber's eye to see. A recipe for deep-fried scorpions. The apparent fact that cracks in breaking glass move at 3,000 miles per hour (a snippet he traces back to *Popular Science* magazine in 1939). That kind of thing.

Zoology, I find, is full of stuff. It has been discovered, for example, that Canadian elk can be divided into two basic personality types – 'bold runners' and 'shy hiders'. In the age of bow and arrow it was the former that fled to safety and the latter that were stalked and killed. Now it's the other way about. In the age of high-powered hunting rifles, it's the runners that end up in the cross-hairs and the wallflowers that survive. For the first time, say scientists at the University of Alberta, we have proof that humans can affect the 'personalities' of the species they hunt. It's *unbelievabilia* because, though I can see it might be true, it is not something I could have imagined. I dip

randomly into a pile of academic papers, and stuff comes out
in handfuls:

Animals are like football teams, having a tendency to esca-
late their conflicts (more fouls, more fights to the death)
when they are closely matched.

Chimpanzees have a formal 'grooming handclasp', which
involves two animals gripping each other's wrists or arms
and using their free hands to groom each other. The grip
varies between communities, which suggests that their social
behaviour may be culturally as well as genetically influenced,
just like our own. Stealing their mates' food, on the other
hand, shows that they have little concept of honour.

Hyenas are determined problem-solvers. Confronted with
steel 'puzzle boxes' packed with meat, they will go on
trying different ways of getting inside until they find the
bolt. Researchers from the University of Michigan say it
proves the advantages of a large brain.

There is a very good reason why ground squirrels habitu-
ally wave their tails in the air. According to the University
of British Columbia, it keeps rattlesnakes away.

Mother goats can recognise their kids' voices even after a
year of separation.

On average every month, pumas in Patagonia leave 232
kilograms of meat per 100 square kilometres to be eaten
by condors and other scavengers. This is 3.1 times more
than wolves leave in Yellowstone National Park.

The Philippine tarsier can communicate at an ultrasound level of up to 91 kHz. Researchers from Humboldt State University say this might represent a 'private channel of communication' inaudible to predators.

The most ecologically distinct mammal in the world – the one with the fewest relatives – is the aardvark.

It is here that *unbelievabilia* begins to nudge up against something more serious. It is not before time. So far the ideas in my head have been like an intellectual construction kit with no assembly instructions. How am I to put all this together into a cogent narrative? My interest in the Somali golden mole has been easier to *feel* than to explain. The 'quest', as I have rather romantically called it, can come across as bizarre or eccentric – pretentious even, a facile and useless specialism like collecting novelty teapots. It is Professor Jonathan Baillie, with a little help from the aardvark, who steers me back to coherence and a sensible appreciation of my own seemingly childish instinct. Jonathan is a fast-talking Canadian zoologist and a world expert on obscure mammals. Thankfully, he can see nothing silly in my mole obsession. He is, after all, surrounded by people fixated on solenodons, echidnas, freshwater dolphins, elephant shrews . . . We meet in a conference room at the Zoological Society of London, just across the road from the zoo. This is his domain, the command centre of a worldwide effort to bring relief to the deserving mammalian poor, the small, the weird and the unheard-of. In 2007, in what might be one of the most powerfully imaginative strokes in the history of conservation, Jonathan founded the EDGE Project. 'Edge', no surprise, is both a word of literal meaning, as in outer margin, and an acronym – *Evolutionarily Distinct and Globally Endangered*. EDGE means the aardvark. It means the woylie, the northern muriqui, the long-footed potoroo, the Dinagat bushy-tailed cloud rat, Perrier's sifaka, the Ethiopian water mouse. It means the Somali golden mole.

Jonathan has no quarrel with the heavily backed and well-publicised campaigns for popular favourites – giant panda, Bengal tiger, snow leopard, black rhino, polar bear, elephant, orang-utan, gorilla and the rest. Their charisma is a powerful tool for attracting public and political support for wildlife. But these animals are not alone in needing help; nor are they uniquely deserving or uniquely important. Some consistency is needed. When human populations are at risk, we do not just rescue celebrities and high-earners. Our concern is for the Common Man, and so it should be for the Common Animal. But of course there is a vital difference. The common animal increasingly is not common at all, and unless we rally to its aid it will quickly decline from scarcity to extreme rarity and oblivion. Countless species already are queuing to cross the Styx like wildebeest at the Mara, and many already are on the other side.

The EDGE Project's selection criteria are somewhat different from the FFI's or WWF's, but the intention – to conserve life and biodiversity – is exactly the same. There are good scientific reasons why the first species on the EDGE list, its top priority, is not a popular favourite but a creature few non-zoologists will have heard of. Attenborough's long-beaked echidna, *Zaglossus attenboroughi*, is classified by the IUCN as critically endangered and, most importantly, it is 'evolutionarily distinct', meaning that it has very few living relatives. Unlike a mouse, therefore, it carries what Jonathan Baillie describes as a 'disproportionate amount of our evolutionary history', having a whole limb of the phylogenetic tree to itself, or sharing it with very few others. It was this kind of unique evolutionary history that EDGE set out to identify and protect – a completely new way of establishing priorities. With its expert staff of phylogenists, ecologists and zoologists, the Zoological Society of London was the ideal platform from which to launch a new worldwide campaign. Working closely with the IUCN *Red List*, and inching their way carefully through the phylogenetic trees, Jonathan and his colleagues devised formulae to score each species for evolutionary distinctiveness and risk of extinction. By aggregating the scores they were then able to produce a table of species ranked in order of need.

It is a very big league of often very small animals. The list of mammals (there is a similar list of amphibians) runs to nearly 4,500, from Attenborough's echidna right down to the gray brocket (a South American deer) at number 4,436. Another 920 are unranked because they are too poorly known to be given a score, and sixty-five are already extinct. The emphasis on evolutionary distinctiveness – what a layman might call *uniqueness* – means that many of these animals combine extreme peculiarities of appearance with oddities of behaviour. All are in some way unique – 'weird and wonderful' is the common expression – and their loss ought to be unthinkable. Alas, at a time when extinction rates are a thousand times higher than the fossil record suggests would be normal, losing a species is all too easily imaginable. Rivet after rivet is popping out of Paul Ehrlich's aeroplane wing, and the risk of a crash is becoming ever more acute. Hard choices have to be made. Not even the wildest fantasist could imagine that all 4,436 species can be rescued. The Icarus principle applies. Flying too high, trying to do what can't be done, is a short cut to disaster. With heads ruling hearts, Jonathan and his team have to concentrate their efforts on the top 100 species in the list.

By coincidence I switch on the radio this morning and hear

Jonathan speaking from the IUCN World Conservation Congress in South Korea, where he has just launched another list. This one, titled *Priceless or Worthless?*, is both simpler and more complicated than EDGE. Simpler, because it concentrates solely on rarity, not evolutionary distinctiveness. More complicated, because it includes species of all kinds – plants, fungi, invertebrates, birds and fish as well as mammals. These are, quite simply, the 100 most critically endangered species in the world, identified through the combined efforts of 8,000 scientists involved in the IUCN Species Survival Commission. As I said earlier, I have had direct experience of the BBC's mania for editorial 'balance': every action or idea, however exemplary, must have someone to talk it down. This morning we get a real corker. Jonathan is 'balanced' by a woman who says it is illogical to regard all these species as deserving of life. If that's the case, she argues, then we should care just as much about the smallpox virus, and about species that are already extinct – *Save Our Dinosaurs*. Jonathan and the interviewer, James Naughtie, somehow manage to avoid the word 'bonkers' or any of its synonyms. No such restraint is shown in my kitchen, where the radio at this time in the morning is used to being shouted at. The key issue, which Jonathan patiently reiterates, is that the great majority of the *PoW* species have been brought low by humans, and in most cases humans could reverse the tide. Unlike the dinosaurs, which died as victims of nature, they are entirely dependent on the goodwill and mercy of humans. What could be our moral case for denying them? Do these animals have the right to exist, or do humans have the right to exterminate them? For non-contrarians, these are not difficult questions to answer.

To hammer the message home, the IUCN also appends a long – a *very* long – list of species that have already disappeared: eight dolorous pages of squint-small print. Seventy-seven of these are mammals, their identities somewhat irritatingly (if I may carp for a moment) obscured by the absence of common names, as if they are of interest only to science professionals. But why should laymen be spared their morsel of grief for the aurochs, the Hispaniolan edible rat, the pig-footed bandicoot, the giant fossa, the Madagascan dwarf hippopotamus, the sea mink, the indefatigable Galapagos mouse, the Jamaican rice rat, the desert bandicoot, the broad-faced potoroo, the bulldog rat . . . ? Wouldn't even a zoologist find it easier to mourn the big-eared hopping-mouse than *Notomys macrotis*? Ironically, one of the most beautiful illustrated books in my possession is *A Gap in Nature*, written by the peerless Tim Flannery with paintings by Peter Schouten. Together they describe and illustrate 103 species of mammals, birds and reptiles that have become extinct since 1500, including some of the most distinctive species ever to have lived. There among others went Steller's sea cow (declared extinct in 1768), the bluebuck (1800), the white-footed rabbit-rat (1845), St Lucy's giant rice-rat (1852), Gould's mouse (1857), the large Palau flying-fox (1874), the Falkland Islands dog (1876), the eastern hare-wallaby (1889), the Santa Cruz tube-nosed fruit-bat (1892), the red gazelle (1894), the long-tailed hopping-mouse (1901), Pemberton's deer-mouse (1931), the desert rat-kangaroo (1935), the thylacine (1936), Toolache wallaby (1939), Caribbean monk seal (1952), lesser bilby (1950s), Ilin Island cloudrunner (1953), Little Swan Island hutia (1955), crescent nailtail wallaby (1956), Bavarian pine vole (1962), greater short-tailed bat (1965), Guam flying fox (1974) . . . And so it will go on until we find a way to stop it.

Twenty-two mammals are cited by *PoW*, of which fifteen also feature in the EDGE top 100 – not only rare, but members of the evolutionary aristocracy. 'Rare' is a word of almost pathetic inadequacy. The northern muriqui woolly monkey is reduced to fewer than 1,000 individuals, the pygmy three-toed sloth to below 500, the vaquita to below 200, Javan rhino to below 100, and Santa Catarina's guinea pig to between forty and sixty – as near to extinction as a living species can get. The idea of fighting for these beleaguered minorities came to Jonathan Baillie while he was studying for a Masters at Yale. During an internship at IUCN he worked on the *Red List* (he is now one of its principal editors) and realised that there were huge numbers of distinct and uniquely wonderful species which hardly anyone knew or minded about. Necessarily he is an optimist. A whole generation of environmentally attuned people had grown up caring about tigers, gorillas, pandas and the rest of the megafaunal pantheon, and Jonathan believed they might now be helped towards a deeper awareness of the full diversity of life, and might even share his sense of urgency about protecting it. The result of that belief is EDGE.

All conservation bodies like to make emotional appeals on behalf of our as-yet-unborn grandchildren. But EDGE is bigger than that. As Jonathan says, the listed species embody a dispro-portionate amount of the world's evolutionary history and, hence, of its biological diversity. This is why they hold out the best hope for the future of life on earth, far beyond the short-term horizon of our grandchildren's grandchildren. For a range of reasons – natural as well as man-made – some of these species will not survive. Given the right kind of help, however, there could be decent chances for the majority. There may be a soft glove of sentiment, but there is a hard fist of science inside it. The planet will go on changing as a perturbed climate inexorably alters the ranges and habitats of species which will have to adapt and evolve to meet new circumstances. The wide biological diversity and varied ecological tolerances of the EDGE species maximise the likelihood that this will happen. The more species, the better.

In one way Jonathan's optimism was fully justified. Before EDGE was launched, sceptics told him not to expect much in the way of interest from an apathetic media. In the age of celebrity, they said, people were not going to care about species they had never heard of. Aardvark, schmaardvark. What does it matter? But the pessimists were wrong. For two whole weeks, Jonathan had to clear his diary for interviews. 'It was just madness,' he now says. 'People loved hearing about these creatures and were really shocked that they were threatened.' Talk, however, runs out of the radio like water. Turn the knob and there it is, a constant stream of catastrophes, good causes and special pleadings. People may be shocked by the rate of extinction, and fascinated by oddities like the poisonous solenodons, but converting interest into action calls for something closer to alchemy. The scale of the challenge is enormous. Against each species in the EDGE top 100 is an assessment of the conservation effort being devoted to it. There are three categories – *active*, *limited* and *none*. For the top five species – Attenborough's long-beaked echidna, the eastern long-beaked echidna, western long-beaked echidna, New Zealand greater short-tailed bat, baiji, all of them critically endangered – the conservation assessments read *none*, *limited*, *none*, *none*, *none*. In all, forty-four of the top 100 have no ongoing conservation of any kind; twenty-two have limited action and only thirty-four are active.

We should remember that the EDGE scores are composites, taking account of both evolutionary distinctiveness and vulnerability. This is why the aardvark – world champion for distinctiveness, but of 'least concern' to the IUCN – is only 313 in the list. Like many of the golden moles it is not often seen but is relatively common in its local habitats. In fact three species of golden mole rank higher than the aardvark. One of them, Marley's golden mole (*Amblysomus marleyi*), is only two places outside the favoured top 100. Juliana's (*Neamblysomus julianae*) comes in at Number 295, and the rough-haired (*Chrysospalax villosus*) at 304. The Somali golden mole, being in IUCN terms 'data deficient', is listed but unranked. Low rankings may be reassuring. It's encouraging for friends of the gray brocket to know that its numbers around the forest margins of Argentina, Bolivia, Brazil, Uruguay and Paraguay are sufficient for it to arouse little or no concern and to be EDGE's lowest ranked species at Number 4,436. But much depends on your standpoint. It's not much consolation for a British wildlife-lover to see the red squirrel down in 4,123rd place as a species of least concern. In Britain it's a goner. Nor is one cheered by the fact that the highest ranked British terrestrial mammal, the common dormouse, *Muscardinus avellanarius*, comes no higher than 840th. It simply reminds us that we have so few species left to care about.

I made some resolutions when I began this book. I would not pretend to knowledge I did not possess (resolution kept); I would not write in anger or deliver homilies (resolutions failed or wavering); and I would not heap opprobrium on men of the past who inhabited a different moral landscape (resolution kept). But a question remains: through what moral prism should we view the behaviour of our own generation? How might we be regarded by generations in the future? Let us not be lured into making false comparisons. It is on the basis of *mens rea* – the guilty mind – that the old-timers may be acquitted. They did not know – indeed, they had no way of knowing – that they might be stripping the planet of life or, through their God-given technological genius, putting an intolerable strain on the climate. They might have been immodest in dealing with their fellow man, but they knelt to God and saw no possibility that they could undo the work of Genesis. This is an opinion that now survives only in the minds of extreme libertarians who would rather boil the oceans than submit to regulation of the free market. Whatever we do, we do it in the full knowledge of its likely consequence. We cannot plead ignorance.

Modern warfare is wholly impersonal. No one sees the whites of their enemies' eyes any more. Remote push-button deaths are a mathematical abstraction, swiftly escalating beyond the point at which the numbers can be visualised. They are always there, a muffled drumbeat in the broadcast news. 'Yesterday in Iraq, Afghanistan, Yemen, Syria, Nigeria, Sudan, Somalia. . . .' Self-censoring news media spare us the kind of film and photography that would rub our noses in the reality of it. The war on wildlife is just as impersonal and even less visible. There is no 'Yesterday in Kenya, Tanzania, Senegal, China, Brazil, Indonesia . . .' Just occasional round-ups of numbers pushed to the brink. Even these are wildly theoretical. To make such a calculation we would have to know how many species there were in the first place, and have some reliable idea of what existed where. This is extremely difficult – one might better say impossible – and it's an uncertainty that gnaws at the soul of everyone seriously involved in conservation. You can't conserve a critically endangered species unless you know where it is. You can't know where it is unless you mount a very expensive – and very likely inconclusive – expedition to find it. The obstacles are financial, logistical, political, physical and technological.

Jonathan Baillie spends more time on management now than he does on field work, but he did lead an expedition in 2007 to search for Attenborough's echidna, now the EDGE list's Number One species. The echidna had a significant advantage over the Somali golden mole in that a specimen, just one, had actually been seen alive, though this was a very long time ago – in 1961, three years before Alberto Simonetta found his owl pellet. The animal was collected 1,600 metres above sea level, on Berg Rara in the Cyclops Mountains of Indonesia. It earns its position at the top of the list by being both extravagantly rare and, as Jonathan puts it, 'one of the most distinct mammals on the planet' – a small, spiny creature that looks superficially like a hedgehog but with a long, bird-like beak. It amazed him that so little effort had been made to find it.

Jonathan Baillie in Papua New Guinea

But the looking is the easy bit. First you have to define the field of search (not especially difficult in the case of an animal seen only once), and then you have to get yourself there. This is difficult in all the ways you would expect – steep, thorny, unmapped and treacherous mountain terrain – and in many ways that you perhaps would not. Particularly in tribal lands, where there are often invisible layers of bureaucracy and a variety of spoken languages, the territory is full of diplomatic tripwires. This was certainly the case with the echidna. 'You have to speak to the Indonesian representatives,' says Jonathan. 'You have to speak to the local forestry representatives, you have to speak to the local tribal leadership, and you have multiple conservation groups that you'll be working with, so you have to meet with the head of each one, and then you have to deal with the military, because they have a base on the side of the mountain.' You then need transport (including, in this case, a boat to reach the coastal villages), guides, interpreters, all the paraphernalia of a scientific expedition. Then it comes down to talking; moving from village to village, each one with a different language and cultural traditions, trawling for news of the species. 'You have to make contact. You can't just walk in and start demanding things. You have to see if they're interested in the species. You get them to tell about their history and the animals and the different things they hunt and how far they go in the forest.' The questioning has to be indirect. You can't just show pictures, or describe an echidna, and say, 'Have you seen one of these?' Leading questions produce biased results. You have to get people to describe all that they've seen, and it takes time. 'There's tons of local knowledge if you sit and listen, and it doesn't always come out right at the very beginning. You have to build trust. You're building also an understanding of all the other species and what their ecological parameters are. And you're looking at opportunity. What's the best and most feasible route to get in and start looking? And obviously listening to where they've seen the animals last.'

With the echidna it worked perfectly. The descriptions were so precise, and so consistent from village to village, that Jonathan was in no doubt. Forty-six years after the first and only sighting, the animals were still there. After that, it was just a question of deciding where to set up camp and where to search. Attenborough's long-beaked echidna sets a number of challenges typical of EDGE species. It is small, nocturnal and confined to hard terrain, in a faraway country with an ambivalent attitude to conservation (the illegal burning of Indonesian forests to make way for palm oil plantations, a catastrophe for orang-utans, is a scandal without parallel). For a small search team working by torchlight, the difficulties are formidable. Diligence alone is never going to be enough. You need the gods to smile, and the best you can do is to be ready to capitalise on any stroke of luck.

For the echidna team, luck came in the form of nose-pokes. This is not the rude awakening it sounds. Echidnas have a very distinctive way of feeding. Their preferred food is worms, which they find by prodding the ground with their long tubular beaks. Hence the nose-pokes, which they leave as evidence of their passing. There, however, the team's luck ran out. The architects of the nose-pokes declined to show themselves, alive or dead. Even so, it was a result. Enough for the *Red List* to declare with confidence that *Zaglossus attenboroughi* still existed, though the impacts of hunting, its restricted range and shrinking habitat meant that it remained critically endangered.

In terms of conservation status, an EDGE listing is a badge of honour. A species may gain enormously from its imprimatur, especially if it is highly ranked. In particular, it significantly improves the chances of funding. There is a real hope that Attenborough's long-beaked echidna will be saved, though the current conservation effort is only 'limited' and there are no guarantees. The sad story of the baiji, *Lipotes vexillifer*, stands as stark evidence of the abyss that stands between recognising a species in need and doing something effective to save it. When EDGE published its first mammals list in 2007, the baiji stood where Attenborough's long-beaked echidna now stands, at Number One. By 2011 it had slipped to fifth, though not because there had been any lessening of risk. On the contrary, by then its chances of survival were vanishingly faint. The alternative common name for *Lipotes vexillifer* is Yangtze River dolphin, which describes it perfectly. It is, or was, a freshwater dolphin found exclusively in the Yangtze, a species of beauty and grace, which, according to legend, was the reincarnation of a drowned princess. The lovely animal and the densely polluted shipping highway, booby-trapped with fishermen's nets, in which it found itself struggling to survive were a gross mismatch, which, without intervention, could have only one outcome. Saving the baiji became an international *cause célèbre*. The tragedy, the *disgrace*, is that it should have succeeded. An appalled witness to the debacle was one of Jonathan Baillie's colleagues at the Institute of Zoology, Dr Sam Turvey. An hour-long conversation with him, I now confess with some embarrassment, fell victim to my incompetence as a sound engineer, producing a fine recording of my questions but nothing at all of his answers. Never mind. Making no attempt to conceal his distress at what he had seen, Sam set it all down in a book with the no-frills title *Witness to Extinction: how we failed to save the Yangtze River dolphin.* Everyone knew what needed to happen. Baijis should be caught alive and transferred away from the unsurvivable river channel to breed *ex situ* in an oxbow lake. But the Swiss-funded rescue plan turned into a rolling farce of obfuscation, procrastination and tragi-comical bungling, exacerbated by a malign Chinese bureaucracy constitutionally unable to match actions to words. In effect, the baiji was talked out of existence. In late 2006, a year before EDGE came into being, Sam Turvey joined an expedition that combed the Yangtze from one end to the other of the baiji's historic range, and back again. Not a single dolphin did they see. The baiji had gone, with the result that – 'evolutionary distinctiveness' meaning what it says – there was nothing like it left anywhere on earth. Biodiversity had just taken an almighty hit.

Sam chides international conservation organisations for failing to put their weight behind the baiji. 'This looks like a good project. Good luck. Let me know how it goes,' was a typical response to his plea for help. It led him to ask a sharp political question. Is extreme vulnerability actually a hindrance to conservation effort? Conservation charities like to be associated with success, and the risk of failure may be a powerful brake on their willingness to intervene. Could it be that some species of extreme rarity were deemed too high a reputational risk even to try to save? 'On the other hand,' he writes, 'organisations which are too timid to put themselves on the line instead only fail passively, and can cover their tracks and justify their inaction without even needing to apply too much spin. Paradoxically, if conservation organisations are run like businesses, then maybe not trying at all might even become the better option.' You could argue that the NGOs had a point. It's no good throwing money at lost causes. But then the baiji turned into a lost cause only because the chance to save it was missed. If the world needed an object lesson in how not to protect a unique species, then this was it. As Sam persuasively argues, the triage system should not be an excuse for inertia. What the world *did* clearly need was a pro-active standard-bearer for threatened wildlife that would identify priorities, draw up action plans, lead and coordinate the efforts to see them through. Out of necessity, EDGE was born.

The questions it raises have been unconsidered on such a scale since the authors of the Old Testament laid down their papyri. We do not know how many species the planet is home to, but we can reasonably suppose that the undiscovered ones are in no better shape than the ones we know about. By some estimates 20 per cent, and by others 30 per cent, of the world's species are imminently threatened with extinction. At least one in five. That would be a shocking rate of attrition if it applied just to individuals. Imagine the outcry in England if, say, one in five dogs or donkeys were condemned to death. Consider the outcry that *did* occur when the government wanted to cull badgers – a policy of extreme prejudice to selected animals but of no threat to the species. We fix on these things because they are within our imaginative grasp. One in five species is way beyond our understanding. As a distinguished scientist will remark in the next chapter: 'We don't know what we are doing.'

As I am writing these words, an email arrives from the Royal Society. Attached to it is a peer-reviewed paper from Stony Brook University, New York, whose title is a frequently asked question: 'How does climate change cause extinction?' You might expect a simple answer. It's a heat-in-the-kitchen thing. Animals will die simply because they are physiologically unable to tolerate soaring temperatures. Look at what happens to humans when they are hit by extreme heat. In the European heatwave of 2003, when even Switzerland scorched at over 40°C, mortuaries ran out of space. Across Europe, perhaps 35,000 people died from the effects of heat. So there you have it. Species that evolved to live in cool conditions will have no answer to the blood-boiling temperatures of the next century.

But of course (even if you accept that climate change will be that rapid and extreme) it's not that straightforward. Nature is a complex weave of interactions and relationships in which causes and effects are rarely direct, and almost never singular. It is generally predicted that the number of extinctions due to climate change over the next hundred years will be measurable in the thousands. But what will kill them, the Stony Brook scientists argue, is not heatstroke but the secondary effects of an altered environment. Some spiny lizards, for example, have become locally extinct because higher temperatures shorten their daytime activity and they don't have time to mate. Other species may die of starvation because they don't have time to feed. Reduced rainfall causing water shortages will have the same effect. So will changes in vegetation, which eliminate food or habitat. So will forest fires. So will rising sea levels caused by melting icecaps. Much more important than all these, however, are relationships with other species. These, too, vary in complexity. If a prey species becomes locally extinct, either by dying or by moving elsewhere, then predators will die out too, and so will the parasites and scavengers that depend on them. Some species may thrive in a warmed climate, or move successfully into new territory, where they may out-compete native species or infect them with diseases (Britain's grey–red squirrel conflict provides the perfect example). Pollinating insects may move off. There will be mismatches in the changed seasonal behaviour of dependent or interdependent species – birds and insects hatching at different times, for example. Any or all of these can throw an ecosystem out of kilter, the loss of one species leading to the loss of others, a gathering avalanche of extinction.

It so happens that these arguments have been made in a paper about climate change. But it doesn't matter how a species dies – through breeding failure, starvation, habitat loss or persecution by humans – a dead species is a dead species and the result is another mortal blow. This is why EDGE matters. Without support, many of the world's most distinctive animals will be gone within a decade. For us, the architects of mayhem, it is make-your-mind-up time. Do we care, or do we not? 'If we do,' says Jonathan Baillie, 'then government, industry and the public should get behind these animals. If not, then society has decided it's happy for much of the world's diversity to go extinct.' The great risk in writing in these terms, quite apart from the displeasing sound of one's own unwonted querulousness, is that people will conclude that we are beyond a tipping point, and that nothing we do now can make a difference. All they can hear is pounding surf on the reefs of doom.

But it's not like that. Just listen. Wherever you are, you won't have to go far to hear birdsong. Even in biologically depleted Britain, gardens teem with life. In every continent at every degree of latitude, in mountains, forests and plains, enthrallingly beautiful creatures still live as they have done for millennia. Like the baiji once was, they are savable. Like the baiji's, their direct ancestral lines are far longer than our own, way older than anything imaginable to the scribes of Genesis. I run my eye once again down the list of threatened mammals, like names on some predictive war memorial, lifted from a nightmare. Sumatran rhinoceros, riverine rabbit, wild Bactrian camel, Asian tapir, Mediterranean monk seal, pygmy hippopotamus, Gilbert's potoroo, dugong, western gorilla, blond titi monkey, Amazonian manatee, Ethiopian water mouse, blue whale, giant golden mole . . . Who would pull the trigger on any of these? Who would pull the trigger on any of the thousands of others? It is no fault of theirs that we regard so many of them as obscure – their obscurity is our ignorance. The first priority is to plug the gaps in our knowledge, which means winning the long-term support of governments. That requires commitment, training, and a concerted effort, which, like nature itself, is blind to political boundaries. Knowledge is all, which is why EDGE has established its own Fellowships and a highly developed training programme whose graduates are already working across the globe. 'So,' says Jonathan Baillie, 'we're trying to build up a generation that can respond, that cares about these species. What we ask them to do is create a blueprint for survival, which is really the initial stages of an action plan, trying to identify what needs to be done. Sometimes we know a species is very scarce, but we don't know what caused the decline. Just understanding the basic threat is sometimes a big step forward.'

Surveying endangered species can have unexpected benefits too. When 'new' species are discovered, it is almost always the result of having looked for something else – an unknown mammal by definition cannot be sought. If you're lucky, you just happen across one. This enticing possibility was in the back of Jonathan Baillie's mind as he hunted for Attenborough's echidna. In such unexplored terrain you literally never know what might turn up, and in no other science is serendipity so important. Often, says Jonathan, discoveries happen when a villager brings out of his house something he has just caught – familiar to him, who might have been stewing it with vegetables all his life, but a novelty to science. Or you might spot something for sale in a market. Down in the basement of the EDGE list, whose denizens are too few or too obscure to rank, is the red gazelle, *Eudorcas rufina*, which, as I mentioned in Chapter Six, is famously known from three male specimens bought in the late nineteenth century at markets in northern Algeria. It has never been seen in the wild, and its provenance is a matter of some doubt. At the other end of the list, in fifteenth place just below the Javan rhinoceros, is the kha-nyou, *Laonastes aenigmamus*, sometimes known as the Laotian rock rat, or rat squirrel. Had it been discovered in the early nineteenth century, it would have been exactly the kind of species at which the learned sceptics in London would have tutted. With the body of a large rat and the tail of a squirrel, it looks exactly like a Phineas T. Barnum stitch-up, classic *unbelievabilia*. Today there may be questions about its phylogeny, but none about its authenticity. The kha-nyou was first discovered in 1996, laid out in a meat market in Laos. Three more specimens were supplied by villagers in 1998, along with some remains in an owl pellet, and others have been found at roadside stalls. It was originally assigned to a whole new family of its own, the *Laonastidae*, but has since been awarded a much more exciting pedigree as the only living representative of the primordial family *Diatomyidae*. All its known relatives exist only as fossils. The kha-nyou is, quite literally, a living fossil, the rodent coelacanth, evolutionary distinctiveness on stilts. If EDGE ever wants an emblematic species, then it need look no further than this.

Thirty-one places below the kha-nyou, at Number 46, is the golden-rumped sengi, also known as the golden-rumped elephant shrew, *Rhynchocyon chrysopygus*, a rabbit-sized native of north-eastern Kenya. (In the interests of *unbelievabilia*, I check to see how many species listed by the IUCN have the word 'golden' in their name. The answer is 295.) In 2010, during an expedition to survey the sengi, an EDGE Fellow, Grace Wambui, spotted another species she could not put a name to. Camera traps revealed an animal about two feet long, with large eyes and ears, spindly legs, wiry tail and a long trunk-like snout. Its upper parts were a grizzled yellow brown, its thighs maroon and lower rump black. It was evidently a giant elephant shrew, but not one of the four species already known to science. (There are also thirteen species of the smaller, soft-furred kind.) Thus did the elephant, aardvark and golden moles acquire a new cousin. It will be a challenge for EDGE to make sure that they hang on to it. The animal's obscurity until now was guaranteed by its confinement to what was in effect a scientific no-go area, dangerously close to the border with violent and lawless Somalia. Now improved security means that loggers are moving in and doing their lethal worst.

Camera-trap photograph of a new species of giant sengi in Boni Forest, Kenya

In 2012, a new monkey – *Cercopithecus lomamiensis*, known by the Congolese as the lesula – was discovered in the Democratic Republic of Congo, a blond-maned species with the long, lugubrious face of a Roman magistrate. New monkeys in Africa are rare. The previous discovery was the kipunji, *Rungwecebus kipunji* (Tanzania, 2003), and the one before that the sun-tailed monkey, *Cercopithecus solatus* (Gabon, 1984). Their hold on life is slippery. The kipunji is hunted for meat and losing habitat to loggers and charcoal burners. The IUCN lists it as critically endangered. The sun-tailed monkey, subject to similar pressures, is listed as vulnerable. So, already, is the lesula. For the conservation community, the ultimate test is to prevent the celebration of discovery from turning immediately into the mourning of loss. Recent history offers some doleful examples. The baiji's ancestors lived on earth well over 100 times longer than humans have been here, but the baiji itself was a stranger to science until 1918, when it had less than ninety years left to live. After the great whales, the fabulous Steller's sea cow was the biggest mammal to survive into the modern historical era. It stretched up to 8 metres from nose to tail, and weighed 10 tonnes. The story has a typically seren-dipitous, almost romantic beginning. In 1741 the naturalist George Steller was shipwrecked on the Commander Islands in the Bering Sea, where he observed these huge animals living in herds around the coast. Peter Schouten's painting in *A Gap in Nature* shows a long, torpedo-shaped animal with an amiable, soppy-dog face and skin like tree-bark. Steller's own description of it, quoted by Tim Flannery in the same book, has a prophetic poignancy:

> They are not in the least afraid of human beings . . . they have an extraordinary love for one another, which extends so far that when one of them was cut into, all the others were intent on rescuing it and keeping it from being pulled ashore by closing a circle around it. Others tried to over-turn the yawl. Some placed themselves on the rope or tried to draw the harpoon out of its body, in which indeed they were successful several times. We also observed that a male two days in a row came to its dead female on the shore and inquired about its condition.

'Not afraid of human beings . . . cut into . . . rope . . . harpoon . . . dead . . .' For hunters looking for meat, oil and skins, the sea cows were as easy a target as their terrestrial namesakes in a field. By 1768, just twenty-seven years after its discovery, the species was extinct. At least we have the small, salutary consolation of knowing that it once existed: we know and can regret our loss. It is fair to assume that in the beleagured forests and poisoned waterways, many others will perish without ever bearing a name. For all the species I have mentioned, I could have substituted a shockingly large number of others. My choices have been arbitrary, but so too are the processes by which we drive animals to the edge. For the baiji, Steller's sea cow, bluebuck, kouprey, Caribbean monk seal, Japanese sea lion, Sardinian pika, and all the others from the obituary pages of the last 500 years, there is no possibility of return. For all those still clinging on, there is hope worth investing in. Habitat protection, translocation and captive breeding can all work to stop the dreaded code-letter E appearing with such awful frequency in the IUCN *Red List*.

Hope lurks in stranger places, too. In 1987 a group in South Africa set out to recover a lost animal through selective breeding. This was the quagga, a sub-species of the plains zebra once common in South Africa but rapidly hunted to extinction. It was distinguished by having the characteristic zebra stripes only on head and neck, with plain brown legs and body. The last surviving individual, a mare, died in Amsterdam Zoo on 12 August 1883. Now, by selectively breeding from a herd of southern plains zebras, the Quagga Project aims to rectify that 'tragic mistake' by retrieving the genes responsible for the animal's unique striping pattern. Optimism is the base metal of all conservation projects, and the Quagga Project leaders serve it by the tonne. 'It is hoped,' they say, 'that if this revival is successful, in due course herds showing the phenotype of the original quagga will again roam the plains of the Karoo.' My desk is made of wood, and I am touching it with both hands.

By comparison with the Frozen Ark Project, however, the quagga recovery programme looks like a throwback to the age in which the animal died. Frozen Ark is a jaw-dropper that would have stretched the imagination even of the quagga's far-sighted contemporary H. G. Wells. It was established at Nottingham University in 1996 with the aim of collecting and freezing DNA samples from as many endangered animals as they could get them from – not as an alternative to saving species in the wild, but as extra insurance in case the conservationists failed. Seldom has humankind placed more faith in its own genius. Frozen Ark is now a consortium of twenty-two world-leading zoos, aquaria, museums and research institutions, all united in faith. Progress in molecular biology, they believe, will mean that 'in the not-distant future' lost animals could be recreated from these frozen cells. So far they hold 48,000 samples from 5,500 species. Multiple practical, ethical and moral issues stand between them and the day the samples might be used to relaunch species from extinction. If they perished originally through habitat loss, then where would they all live? But never mind. I'm touching wood again. Could it be that one bright morning in a faraway spring, laboratories will fling wide their doors and out will troop all the lost denizens of the desecrated ark? It takes some believing, but hope at least is not extinct.

One species I know they cannot have sampled is the Somali golden mole. Even now I face difficulty in explaining my fascination with it. The question comes in varying forms, but in essence it is always the same. 'Why is it so important?' My answers vary from the flippant – some people study cheese labels; I'm into owl pellets – to the feebly honest. It's not at all important. That's the point. In its very obscurity the mole stands as a symbol for the whole unsung, unheard-of majority of mammalian life. Apart from which, the pursuit of such a rare creature is terrific fun. In my pocket are two tickets to Florence, an address and a telephone number.

CHAPTER THIRTEEN

The Mole

Fifteen minutes from the centre of Florence, at the bottom of a leafy suburban street, the taxi drops us off, Caroline and me, in front of a green wooden gate. On the gate are an entry-phone and two bell-pushes. I do as I have been instructed, and press them both. Already, at ten in the morning, the sun is burning the top of my head and adding to the nervous sweat. I have not slept well, my mind churning with *what-ifs*. Florence in September is glorious, but it's a long way for a wild-goose chase. What if he's not at home? What if the mole really has been lost, as I have been warned? What if . . . ? But the entry-phone buzzes, the gate clicks open and we find ourselves in a steep overgrown garden, vigorously unruly, of the kind that seems not so much cultivated as *tamed* – an amiable contest between man and nature that has resulted in an honourable draw. To our right as we climb the curving stone steps I notice some old cartwheels and a bicycle; ahead of us, glimpsed through the foliage, the outline of a house. At the top of the steps, on a sunny terrace with a greyhound at his feet, stands a small elderly man wearing a khaki bush-shirt, crumpled clay-coloured trousers and a pair of sandals that look as if it they might have walked here from Africa. He holds out his hand and leads us into the curtained interior. As our eyes adjust to the gloom, it becomes apparent that the house is as much a museum as a home. Professor Simonetta had it built in 1966, but I would have been two centuries out if I'd tried to guess its age. Generations of gilt-framed ancestors peer down from the walls on what looks like a film set of historic clutter. English silver, an inkstand with quills, cases of antiquarian books. The steep gradient means that the front door is on the upper floor. When we go downstairs later to look for wine (he knows there is some but can't remember what) he shows me cabinets full of children's books and toys – cars, tanks, aeroplanes, several wars' worth of soldiers, dolls and puppets stretching back to the eighteenth century. He is not, he insists, a collector. Merely an accumulator. This is all stuff handed down through his and his late wife's families.

The professor is eighty-two, but retired only in the sense that he no longer teaches. He has four papers awaiting publication and a book on evolution on the way. It amuses him that I have come so far to see so little, but then amusement seems to be his speciality. The taped conversations, spread over two days, are full of sentences dissolving into laughter. He walks with a stick but has a filing-cabinet memory and a mind like a steel trap – nothing escapes him, and careless questions are biffed straight back. *'WHAT?'* Within moments it becomes clear that my preconceived idea of him has been a hopeless miscalculation. In my imagination he was an obscure researcher whose career peaked serendipitously in 1964 with the accidental discovery of an owl pellet. In reality, *Calcochloris tytonis* was little more than a briefly amusing footnote in a long and distinguished career that raised him from a seven-year-old bug-hunter to the highly esteemed Professor of Zoology at the University of Florence. So many species have been named after him that even his formidable memory is unequal to the task. 'Well, I can't remember all of them,' he says, 'but one or two species of grasshoppers. At least one lizard. I have, I think, a snake.' Back in England, poring over the textbooks, I manage to track down eight. They include a marine worm, three dung beetles, a grasshopper and a praying mantis. I cannot identify the snake, but am amply compensated by the Coastal rock gecko (*Pristurus simonettai*) and, best of all, Simonetta's writhing skink (*Lygosoma simonettai*).

His curriculum vitae runs to twenty-five pages and lists 280 publications. Some of these, like 'On the distribution and significance of the Paratympanic organ', are academic papers on zoological minutiae far beyond the audible range of laymen (the 'Paratympanic organ', to save you looking it up, is 'a small sensory organ in the middle ear of birds'). There is plenty of stuff, too, on the classification of fossils, the mammals of Somalia, the skull of the dik-dik, 'the myth of objective taxonomy and cladism', loads of grist for the zoological mill. But there are other things, too, that I would never have suspected (or at least would not have expected until I stepped inside his house). 'Some hypotheses on the military and political structures of the Indo-Greek Kingdom'. Essays on the coins of the ancient world. Works in preparation include not only 'The skull morphology of phreatic fish', but also 'A guide to the Parthian Coinage (with a description of the author's collection)'. The width and depth of his focus seem infinite; his interest inexhaustible. And, of course, catalogued as Number 63, there is the paper that has brought me here: 'A new Golden mole from Somalia with an appendix on the taxonomy of the family Chrysochloridae'.

Zoology is a peculiar discipline calling for an improbable combination of cerebral, psychological and physical skills. It's not enough to be an adventurer. It's not enough to have an enquiring mind. It's not enough to have mental stamina and an easy command of minutely nuanced detail. You must have them all. And you must have them all in abundance. The professor's great uncle – his grandmother's brother – was murdered by *shifta* during an exploration of the Omo valley in southern Ethiopia, unruly neighbour of the even unrulier Somalia. His own expeditions to Somalia, Afghanistan and Congo-Zaire cannot have been without danger; and yet the gung-ho zoological commando needs a steady, counter-balancing alter ego who is as adept at the microscope as he is in digging out a stranded Land Rover. It's hard to imagine all this in the professor now, yet his stories have more than a distant echo of Gordon-Cumming or Selous. And we don't have to imagine them. The expeditions are recorded on film. Would we like to see them? he wonders.

It means moving to the dining room. There is a long cluttered table, a dresser crammed with glass and china, and a cushioned bench on which we have to wedge half a buttock and twist our heads as he loads the cassettes. No film was made of his very first expedition to Somalia in 1959, the last year of the Italian administration, when he was a twenty-nine-year-old junior lecturer at the University of Florence. 'We thought it was perhaps the last opportunity to collect animals in Somalia before the administration gave way,' he says. 'It was just a small expedition that lasted two months, but we were very lucky and collected a lot of good things.' It was enough to persuade Italy's National Research Council, the *Consiglio Nazionale delle Ricerche*, to fund a second expedition in 1962. This is the subject of the first video we see, which the professor shot originally on Super8 cine film, with his own commentary in Italian. The swashbuckler and the master of detail merge into a single obsessive record-keeper. Nothing is omitted. We see the gear being stowed aboard ship at Genoa. We watch the banks of the Suez Canal slide slowly past, and then the roll and yaw of the Land Rover as it confronts the Somali interior. And then at last we get down to zoology. There is the young professor himself, dark-haired and handsome, cleaning the skull of a mouse; a colleague stripping a snake; the professor again, skinning a Grant's gazelle. The flayed carcass will be left outside the tent for scavengers to strip before the bones are brought back to Florence. There are some winsome baby genets, and guinea fowl hunted for the pot, which have to be caught alive. The Somali support crew are Muslims and will only eat birds or animals that have been slaughtered by having their throats cut. While we watch shots of a craftsman carving white soapstone, the professor suddenly produces a jug made from the same material, as if the film were transcending time and space and reaching out into the room. Yet the most intriguing shot is so brief that we almost miss it – a swift pan across the facade of a house in Giohar. It is gone so quickly that I have no time to notice, let alone record, any detail beyond a sense of isolation and a shading of trees. This was the house they used as their operational base, the one to which they will return two years later in the crucial year of 1964. It has an outbuilding within which stands a disused oven, and in the oven roosts a family of hungry barn owls. This is the very shrine; the last and only known resting place of *Calcochloris tytonis*.

Disappointingly the film of 1964 makes no reference to the golden mole. We are shown turtles, naked mole rats, vipers, Marabou storks, baboons, egrets. We see dik-diks being caught like rabbits in a long-net, and a dead lioness of a Somali sub-species, collected (i.e. shot) for the Natural History Museum in Florence. It is scenes like this, the obvious inhospitality of the terrain, the careful conservation of water from the roof of the tent, that seem somehow to close the circle, to call upon the spirit of Selous and Gordon-Cumming, brother-adventurers joining hands across the centuries. Fifty years ago, however, the emphasis was already shifting from explorations of the infinite to inventories of the vulnerable. The professor agrees with Jonathan Baillie. Before you can decide what to conserve, he says, you have to find out what is there. The films themselves have lain on shelves, filed away not just as memorabilia but as part of a scientific record. He last watched them, he says, in 1982.

For seekers of small mammals, few things are more propitious than an owl pellet. 'It is a sort of mine for nocturnal and small animals,' the professor says. 'It is always good to collect these, because you have a very complete sample of the fauna.' He laughs at the memory. 'When you find pellets from owls, you always collect them.' Although, sadly for mc, he neglected to film himself doing it, he shovelled out the entire contents of the oven and brought them back to Florence. 'Ninety-nine per cent of it,' he says, 'was – what is it called in English? A sort of mole-rat without hair?' *Naked mole-rat, I wonder?* 'Ah yes. And then there was that one!'

That one! The rarest mammal on earth! But the Somali golden mole, it turns out, is not the only new species he has discovered. He reckons there are at least twenty. Always, the discoveries are serendipitous. For a while, around 1980, he taught zoology at the agricultural university in Somalia. While there he found two specimens of what his colleague Benedetto Lanza would later identify as a new species of lizard. One was discovered among the university's poorly kept collection of skins; the other, more recently alive, was delivered to him in two pieces. A student had gone to pick up a book she had left in the sun, and found a lizard basking on the open pages. 'She was afraid,' says the professor. 'She closed the book like *that* [he claps his hands] and cut the lizard in half.'

The real skill, of course, lies not so much in finding things but in realising that they are new. 'The important thing is the study of the collections, not the collecting itself,' he says. 'Anyone can do the collecting.' Well, as we shall see, that is true up to a point. In trying to find living examples of the Somali golden mole, the professor enlisted local children, who were promised a Somali shilling for every specimen they found. This is usually a reliable method, though in this case they drew a blank. It was no particular surprise. Golden moles live underground, leave little trace on the surface, and make their nests in burrows under bushes which are very difficult to find. 'Unless they are locally common,' says the professor, 'to meet one on the surface is exceptional.' It dawns on me that some of these Giohar Irregulars would have been scarcely younger than Alberto Simonetta was himself when he was admitted to university at the precociously early age of sixteen. He was still precocious fourteen years later when, at thirty, he won a scholarship from the National Academy of Sciences and was let loose on one of the world's most important fossil collections at the Smithsonian Institution in Washington. These were finds from the famous Burgess Shale fossil fields, discovered in 1909 by the American palaeontologist Charles Doolittle Walcott high in the Canadian Rockies. Whole books have been written about the Burgess Shale, whose tens of thousands of 500-million-year-old fossils contain a wider diversity of life than may be found in the oceans of the twenty-first century, many of them unlike any living animals. Not even their discoverer, Walcott, who died in 1927, ever fully understood what they showed. In classifying and describing them in the early 1960s, the professor now admits that he made mistakes. These are seized upon by the pugnacious American biologist, the late Stephen Jay Gould, in his book on the Burgess Shale, *Wonderful Life,* published in 1989. But Gould adds a generous footnote:

'He alone, after Walcott and before Whittington, attempted a comprehensive program of revision for Burgess arthropods . . . he also provided substantial improvements upon several earlier studies, and through his comprehensive efforts reminded pale-ontologists about the richness of the Burgess Shale.' The professor, I realise, is like the tip of an iceberg, the visible mani-festation of an unsuspected life. I had known something of the Burgess Shale before I flew out to meet him – indeed, I had read Stephen Jay Gould's book – but my surfing of the Internet had told me nothing of his contribution to its study. What other surprises might he have in store?

The first is lunch. Waving away Caroline's offer of help, he potters off to the kitchen and returns with ear-shaped pasta, *orecchiette*, floating in a delicate chicken broth. This is followed by a dish of cold beef and chicken served with *salsa verde* and salad; then *sbrisolona*, a sweet crunchy tart flavoured with lemon and almonds. He seems surprised when we pat our stomachs and decline fruit and cheese. This courtly old gentleman passing the dishes seems so different, evolutionarily distinct almost, from the young Simonetta of the films, the adventurer who provisioned his colleagues with gun and knife. But he makes a perfect fit with the distinguished trustee of Italian national parks, the eminent author of papers and books, and the holder of one of his country's most prestigious chairs in zoology. The physical energy of the young man who believed *everything* should be collected – 'Because perhaps no one will ever be there again. Or perhaps people will go there after lots of years and things may have completely changed' – has ceded to an intellectual energy of daunting speed and voracity. He may not be able to remember how many species have been named for him; other-wise everything else races out on a synaptic super-highway that seems to have infinite capacity for names, dates and numbers.

He tells of fresh whale skulls washed up on beaches but belonging to a species nobody has ever seen alive. Of the extinct dwarf emu, of which the number in museums exceeds the number of specimens collected in the wild. Of the muddling by Linnaeus of two different kinds of gibbon. Of a bird, the greater honeyguide (*Indicator indicator*), which eats beeswax. It is a story that particularly delights him. Scientists discovered not only that the bacteria in the birds' gut produce an enzyme that digests the wax – a fact that might otherwise be filed under Just Fancy That – but that the enzyme will also attack the wax covering of tuberculosis bacteria, which exponentially increases their vulnerability to antibiotics. *Eureka!* you might think. The trouble, says the professor, is that these potentially useful bacteria live *only* in the gut of the honeyguide, a parasitic species which, cuckoolike, lays its eggs in the nests of other birds, a fact that seriously complicates the problem of reproducing the enzyme for medical use. Nevertheless, it's a discovery that points yet again to the existence in nature of substances of immense potential usefulness to humans. The professor tells the story in answer to a question – the same old question that everyone always asks – about the point of species conservation. It is precisely because we don't understand their value, he says, that we need to preserve them. At the moment, as species slip away, we have no idea what we might be losing. 'We don't know what we're doing,' he says simply. The fact is, we never have.

He speaks, too, of what he calls the *original* golden mole, the first one to be discovered and described, popularly known in English as the Cape golden mole. 'It has a strange story in the name,' he says. 'Because it is called *Chrysochloris asiatica* and there are no Asiatic golden moles at all.' It turns out to have been a mistake made by Linnaeus himself, who first noted the species in 1758, the consequence of a simple handling error.

The specimen arrived among a job-lot of species that had been collected by one of his pupils in China. Unfortunately for taxonomic and phylogenetic accuracy, the ship on its way home made a call at Cape Town . . .

We arrange to meet again at 10 a.m. on the following day at La Specola, the museum of zoology and natural history, in Via Romana, where he promises to have something of interest to show me. Of course, I know now what it must be, and I am resolved for this one day to become a diarist, to record every detail of this climactic morning. Waking at seven thirty, I draw back the curtains and step out on to the balcony of our hotel room overlooking the Arno. The view is astonishing. Almost directly opposite, across the river, is the fourteenth-century Porta San Niccolo, and high above it, already bustling, Piazzale Michelangelo. A crest of cypresses along the ridge-top creates the impression of a sleeping dog with its hackles raised. Only a tiny trickle of water is coming over the Pescaia di San Niccolo weir, the slow tranquillity of the water in contrast to the traffic teeming along the Lungarno Serristori and Lungarno della Zecca Vecchia. Bells speak from the heart of the old city; sirens of ambulances from the clogged arteries of the new. I even record my breakfast: scrambled eggs, bacon, coffee. And what I am wearing: stone-coloured chinos, trekking sandals, a blue and white seersucker shirt. In a canvas shoulder bag I have notebook and pen, voice recorder and compact digital camera. I cannot remember when a day seemed more portentous.

Already feeling the heat, we make our way along the Lungarno delle Grazie towards the Uffizi and the Ponte Vecchio. Along the narrow pavements we are squeezed between ancient walls and the massed ranks of parked motor-scooters. Their windshields are like translucent wings, a grounded swarm of flying ants. Near the Uffizi I collide with a bollard and bruise my leg. The street vendors are out: garish paintings, leather goods, jewellery, toys. The Ponte Vecchio now is a decorous place with its chi-chi art dealers and jewellery shops, a far cry from its reeking medieval origins when it housed the city's butchers. Across the bridge we head along the Via de Guicciardini towards the Palazzo Pitti, stopping to scan restaurant menus for the celebratory lunch we'll have when the morning's business is over. The Palazzo Pitti bankrupted the banker who began it in 1457. Not so the Medici family, in whose hands it would become a monument to wealth, influence and ostentation. This was the seat of their almighty power, and they meant no one to forget it. Already this morning visitors are beginning the long trek through its galleries, voices hushed as in a place of worship. Our destination now is almost within sight. Beyond the Palazzo the opulence drains away into the nondescript Piazza de San Felice, where a police roadblock is causing chaos. Robert and Elizabeth Browning occupied rooms here from 1847 to 1861 (they are now owned by Eton College and available to rent through the Landmark Trust), though the noise and fumes of the traffic are today a pretty strong antidote to poetic musings. A twenty-minute stroll has transported us through five and a half centuries of human endeavour in which each of our signal virtues – imagination, creativity, generosity – has met its antithesis. We are early, so kill time with industrial-strength espressos at a pavement cafe, where we sit wreathed in carbon monoxide.

La Specola is only a few yards further on, but even so it is not easy to spot. There is a modest signboard and an entrance that could take lessons in grandeur from a stationery depot. I console myself with the thought that great searches often end in unexpected places – indeed, it's the obscure corners that most excite the diligent searcher. But then, La Specola is hardly an obscure corner. Even now, despite the ticket office and the sign outside, I wonder if we are in the right place. Further up the street, perhaps . . . I check my watch. The professor must have been checking his, too. We have been watching the Via Romana but he appears behind us from somewhere inside the building, the rapping of his stick ticking down the last few seconds to the appointed hour of ten. He has travelled by bus – not a feat to compare with crossing Somalia in a Land Rover, but nevertheless a considerable effort for an elderly man who walks with a stick. I want to tell him how grateful I am, but he is already bustling away towards the staircase. As we ascend, I notice that he is still wearing his sandals, but now with a blue plaster on one of his toes. It's an odd thing to notice, and an even odder one to write down, but Florence does strange things to the mind. There is so much grandiloquence, so many monuments to

The man who found the mole – Professor Alberto Simonetta (left) with the author at La Specola

wealth. Even in the glorification of the Christian god, there are so many declarations of temporal power that it takes a plaster on a toe to remind us of how frail we really are. The museum of La Specola is, in the true sense of the word, awesome. You don't have to care how nature works. You might, like Richard Owen, see the hand of a creator. You might, like his opponent Thomas Huxley, or like modern Darwinians and Dawkinsites, see the mysterious loveliness of rational science. It doesn't matter. Faced with the architectural and artistic glories of Florence, it would take a monstrous ego not to feel small. Faced with the miracles within La Specola, even a monstrously egotistical Medici would know humility.

We are met by the curator of mammals, Paolo Agnelli, who will lead us on a tour of the galleries. One of the first rooms through which we pass is the very grand Tribune of Galileo, which was built in 1841 originally to display scientific instruments of Galileo and others, all now removed to the Museum of the History of Science, the Museo Galileo, in the Piazza dei Giudici near the Uffizi (if you want proof of human genius, then here is a very good place to start). La Specola itself was founded in 1775, the first scientific museum open to the public in Europe, beating London's Natural History Museum by 106 years. With life for once imitating art, its early collections of fossils, animals, minerals and plants depended heavily on the magpie tendencies of the Medici. It still feels like a palace treasury.

Several times in my life I have tried to take an interest in geology, and every time I have failed. Not this time. My recorded comments as I'm led around are borderline embarrassing:

Extraordinary! Extraordinary . . . It's amazing. I don't know what all these things are. Basalt, I think. Pink rock from Elba.

It looks like you ought to be able to eat it. They look like they're made of sugar, some of them. They are pink, bronze, black, purple, blood red, opalescent green. Something looks like a lump of frozen seawater. Something else looks like it's been carved out of coconut. Another one looks like flakes of chocolate. And others look like coloured ice, like extravagant puddings . . .

On and on I go, my inarticulacy more articulate in its way than any well-worded scientific analysis. For a moment I've forgotten the mole; forgotten what has brought me here. The weary adult is blown away by his inner child. Who would *believe* such stuff? But already we're moving on, from solid rock into the primitive stirrings of arthropodic life. It is like another compartment of the same multicoloured jewel-case. There are beetles, leaf insects, stick insects, bees . . . On the recording machine I hear what I missed at the time – Caroline and the professor chatting about the fur of golden moles. It is only the Cape species, the misnamed *Chrysochloris asiatica*, he says, that has the famous iridescence described by the *British Cyclopaedia of Natural History* in 1836. I hear myself struggling to catch up, still gabbling into my microphone.

We're now into spider crabs and whatnot. Hermit crabs. An enormous brown crab the size of a small dog. Spotted crabs, lobsters, Norway lobster, crayfish . . . Tape worms. My god! A roomful of intestinal parasites . . .

I realise I am being rude, neglecting the professor, dawdling like an uncooperative child, unable to tear myself from the exhibits. Some mammals are coming up now, and the time-lines suddenly converge. The professor is pointing out some specimens from the very same expeditions that we saw on the films. 'These are dik-diks,' he is saying. 'Guenther's dik-diks.' There are also two larger antelopes, gerenuk (*Litocranius walleri*) and dibatag (*Ammodorcus clarkei*, or Clarke's gazelle). Caroline wants to know if he shot and skinned them himself. 'The big ones yes, certainly.' How very different is this from the Natural History Museum in London, which could not identify specimens shot by Frederick Selous. Here it is like touring the exhibition with Selous himself. There is another difference, too. In London the stuffed specimens are kept as bygones, like a museum within a museum, incidental to its higher purpose. In Florence they are the heart and soul of the place. The professor points to a Grant's gazelle. That, too, came from his time in Somalia. So did a pair of mongooses; and – look! – here are the same little genets we saw on the film. And something I'd never heard of – 'a rare sort of thing', as the professor puts it – a Speke's pectinator (*Pectinator spekei*), named after the English explorer John Hanning Speke, famous for his early explorations of Somalia and his quest for the source of the Nile. The professor surges onward past rabbits, hares, porcupines, flying squirrels, dozens of squirrel-like things that I can't put a name to. Then the ungulates – vicuna, muntjac, Chinese water deer, llama, reindeer, red deer . . . Primates – baboon, mandrills, monkeys, macaques, gibbons, chimpanzee, orang-utan, gorilla . . . A thylacine!

'Yes,' says the professor. 'We have two of them.'

I hear myself remark: 'A lot of people in Tasmania think they're still alive.'

'Well,' he says, 'I hope so.' The whole of creation, or so it seems, is flashing past at the speed of a hurrying eighty-two-year-old with a stick. My voice on the recorder struggles to keep up. *Cuscus, lots of lemurs, aye-aye, more lemurs, tamarin monkey, pangolin . . . Ah!* A brief pause, three taps of the stick, then the professor's voice breaks in to explain the sudden silence. 'These are golden moles.' *Ah!* There are two. One of them nominally is the same as the one on my mobile phone that I photographed in London – the giant golden mole, *Chrysospalax trevelyani* – but this one has been much more carefully stuffed and mounted, so that it looks like a real animal instead of a novelty slipper. The other one, tiny by comparison, is the hottentot golden mole, *Amblysomus hottentotus*, about the size of an English breakfast sausage.

Then we are off again, into the birds of Italy, taped birdsong playing in the background. *Grebes, flamingo, spoonbill, raven, crows, jay, cuckoo, hoopoe, various falcons, peregrine . . . Now we have crocodiles. Alligators. More birds. Pigeons, peacock, nests, eggs . . . Extinct birds. The dwarf emu, the great auk, the passenger pigeon.* This last is one of the dark miracles of extinction. Birds generally lie outside the scope of this book, but the story of the passenger pigeon is too gross to overlook. This is, or was, not just any old species. In the nineteenth century it was the most numerous bird on the planet. A native of northern America, it travelled at high speed – up to 100 kilometres per hour – in flocks of near cosmic size. In 1813 John James Audubon calculated that one such flock contained more than a billion birds, blotting out the sun in an avian eclipse 55 miles long. Sam Turvey in *Witness to Extinction* mentions flocks that stretched for 300 miles and were probably 3.5 billion strong. Their droppings, says Tim Flannery in *A Gap in Nature*, 'fell like snow'. Not any more. Passenger pigeons were hunted with such incontinent voracity that by the 1870s the great flocks were a thing of the past, and a species that had once accounted for 40 per cent of all the birds in North America was spiralling like flying herring into freefall. The very last wild individual was shot by a fourteen-year-old boy at Sargenta, Ohio, in March 1900. The clock had just one more tick to make. A captive bird survived at Cincinatti Zoo until 1 p.m. on 1 September 1914, when it keeled over and took the entire species with it. No wonder the professor stops in front of the specimen and gives us time to ponder.

Then he is off again, past the penguins, the ostriches and owls, pausing briefly by *Titus alba*, the barn owl, whose taste for moles began the whole story. I am noting things almost at random. *Two giant Galapagos tortoises. A Nile crocodile. A leatherback turtle. Snakes. A boa constrictor. A python. But what is this?* I am looking at a peculiarly primitive-looking lizard, like a scaled-down killer from *Jurassic Park*. 'It's a tuatara,' says the professor. 'A sphenodon. It is a unique species from New Zealand, which is practically identical with Jurassic animals.' *So it's a sort of living fossil?* 'Yes, it is.' On again, into fish. *Dorado, sharks, Dover sole, herring, pilchard, tiger shark . . .*

Even though I know it must be imminent, I'm not prepared for what comes next. It is, so far as I know, unique – the thing La Specola is best known for, recommended in all the guide-books (though with the caveat that a strong stomach might be needed). The immediate impression is of a vast butcher's shop, slithering with offal and piled with darkening joints of meat. Truly it is a thing of awesome artistic and technical brilliance; almost impossible to believe it is more than 200 years old, so recently alive does it look. There is not, there cannot be, anything like it in the world. The offal, the bones, the brains, the meat, all of it is human, but modelled in wax. But these are not mere likenesses. They are *facsimiles*. The weight, colour and texture of human tissue are exactly as they were in the dissections they so carefully replicate. The stumps of sawn-off thighs on a pregnant torso look disturbingly ready for the carving knife. The spilling intestines compel you instinctively to cover your nose. The collection fills ten rooms, and there is no part of the human body that is not stripped out for inspection. There are deconstructed heads, faces, limbs, torsos, wombs. The primary purpose was educational, to give medical students the benefits of human dissections without needing actual cadavers – a more sophisticated approach than the English habit of grave-robbing. But art, too, exerts its influence, most obviously in the 'anatomical Venuses', lifelike, erotically posed figures of naked young women spatchcocked with their innards hanging out. These reputedly were much to the taste of the Marquis de Sade.

Beyond this waxen charnel house we come to similar models of dissected animals – sheep, chicken, dog, cat, tortoise – but the professor is picking up speed again, making for the stairs. On the ground floor is a hall of animal skulls and skeletons, not open to the public today but opened specially for our enjoyment. The professor is heading for the whale he told us about yesterday, the one whose skull had washed up on beaches but no one had seen alive.

It is like a gallery of classical sculpture, a display of power and beauty that draws the eye over every plane and curve; nature as art. Overnight, the professor has lent us a copy of his book, *Short History of Biology: from the Origins to the 20th Century*, and in it Caroline has found a quotation from Aristotle:

'. . . So we must, without disgust, begin the study of animals, as in every one of them there appears the beauty of nature, built as they are by nature itself so that nothing is random, but everything is for a purpose, and the purpose for which they are made takes the place that beauty has in a work of art.'

Perhaps I am not the only one whose mind fills suddenly with wordless abstractions. The curator, Paolo Agnelli, puts on a cassette of Mozart, evidence of a sensibility that transcends the ordering of bones. I remember the professor, earlier in the day, reaching into his bag. 'I have taken the liberty,' he had said, and showed Caroline a photograph album of his late wife, Stefania. She was a woman of striking beauty who travelled with him on many of his expeditions. A woman, he tells us needlessly – we can see it in her face – of powerful intellect and forceful personality. The album is simultaneously a purely physical thing, a chemical record of light and shade, and a deeply personal work of art. I am reminded, as I often am, of the separate compartments into which art and science were corralled by the designers of my grammar school education in the early 1960s, as if emotion had no place in the one, and reason no place in the other. It still makes me angry.

Once again in the gallery of bones I have time only for a fleeting record of what I see. *Skulls and whole skeletons of, I don't know what. I think that's a horse. There's an ostrich. A Sumatran rhino. An elephant skull. All kinds of horned animals. Sets of horns. An Indian elephant. A dromedary. A giraffe. Wild boar.* 'This is the whale,' the professor says. 'This was the second specimen discovered. Now there are eight.' In fact, Paolo Agnelli now tells us, there have been several sightings of the living animal itself, the Indo-Pacific beaked whale, *Indopacetus pacificus*. But for some years after it was discovered in 1955 this was the only known evidence of the species since the first skull was collected in Queensland, Australia, in 1882. It is a typical story of sadness and serendipity. The 5-metre-long whale was stranded near

Danane, Somalia, in 1955, whence it was hauled off by local fishermen to be turned into oil and fertiliser. All that remained of it after processing was the skull and mandible we are now looking at.

But, of course, this is not the most important mandible of the day. The moment has come. Leaning on his stick, the professor leads us up the ancient stone staircase, past the public rooms and into Paolo Agnelli's office, where we are ushered to a table in an ante-room. There are glass cases filled with animal skulls, and another boxful on the table. Rhino and goat heads, one dated 1897, are mounted high up on the wall; and there is another ante-room beyond the first, darkened and smelling of insecticide. Groping my way through the gloom, I find the room is packed with stuffed marine mammals in glass cases. Back in the light I return to the table, on which stands a microscope and an empty Petri dish. No trumpet sounds. No drums roll. There is no swelling of strings or cathedral choirs. The only soundtrack to the climactic event, the end of the quest, is my own voice droning on the recording machine.

Paolo puts down a tiny glass phial, about the size of a baby aspirin container. It is packed with cotton wool. Under the cotton wool is an even tinier container, thinner than a thimble. Paolo opens it. And there it is! Calcochloris tytonis!

Two other voices now intrude. The professor's: 'This is all!' And Caroline's, a confidential whisper: 'You're a bit sweaty!' Excitement is dripping off the end of my nose. It is a moment I have been anticipating for months, and yet I find myself strangely unprepared. Of course it was going to be small! I knew that. It was in an owl pellet. But it is so extremely, utterly

minuscule, so completely without consequence, that I can't
believe even a keen young zoologist would have given it a second
glance. 'It's a question of practice,' says the professor when I
ask him how he knew it was something special. When he found
it, the fragment – mandible and part of the ear assembly – was
still intact. Over time, and in handling, it has disassembled into
three separate tiny pieces. With the naked eye – with *my* naked
eye – they make no sense. Paolo puts them in the Petri dish and
places them under the microscope, which he has set to a magni-
fication of six. I ask him if anyone else has ever asked to see
them.

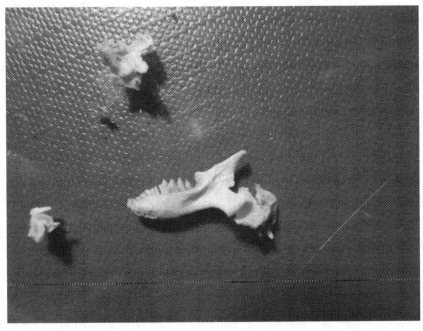

Golden discovery. The jawbone, teeth and middle ear bones of the
Somali golden mole. The animal was eaten by an owl in 1964

'No one for the last twenty years,' he says. And who was the
last? Was it the professor himself? 'Probably.' We all laugh, the
tiny mole's entire circle of friends. Through the microscope,

it looks enormous – *like the jawbone of a whale*, my commentary says, though it's nothing like a whale at all. But it's a predator's jaw, and these are predator's teeth, once eager for insect and worm. All the same, when I try to flesh out the animal in my head the picture has no definition. For months I have carried the mole as an *idea*, but this physical reality, these earthly remains, are too cryptic for my unschooled mind. The professor hands me a pair of thin pointed tweezers so that I can turn the pieces and see them from every angle. This is an extraordinary privilege, a compliment to a competency I do not have. At the Natural History Museum in London, quite reasonably I was allowed to touch nothing. Yet now, at the oldest scientific museum in Europe, this unique and fragile relic is at the mercy of my probing. The largest fragment, the mandible, is a fraction over a centimetre long. The smallest, the malleus, is about the size of a grain of rice. Incredibly, when the professor sifted all the detritus from the Giohar oven, it was this negligible speck that told him it was something unique.

Informed by his earlier experience in South Africa, he had realised at once that it was a golden mole. 'It is very easy to recognise a golden mole fragment,' he says. 'From the teeth, from the shape of the mandible, from the ear. The shape of the ear is very diagnostic.' But there was an anomaly, a peculiarity of latitude. No one had ever seen a golden mole north of the equator before. The professor took photographs, recorded measurements, made drawings and compared his fragments with the equivalent parts of other moles. The suspicion hardened into a certainty. It was typically a golden mole, but different from all others. The professor initially placed his new species, *tytonis*, in the genus *Amblysomus*, in kinship with the fynbos, hottentot, Marley's, robust and highveld golden moles, but – as evolution itself evolved – it was later reassigned to the *Chrysochloridae*, with the Congo and yellow species. It may stay for ever in the bosom of its new family, or it may move on again. With no evidence beyond the Petri dish, the IUCN *Red List* has no option but to classify the Somali golden mole as data deficient and *incertae sedis* – of uncertain placement in the taxonomic tree.

Does it matter? Not really. Nomenclature is an academic diversion, a kind of hobby science that keeps zoologists amused and imposes a pretended order on the chaos of evolution. It gives us a way of knowing what we have, and what we stand to lose, but it has no currency in forest or field, where animals evolve with no reference to their man-given identities. *Calcochloris tytonis* may be 'related' to *Calcochloris leucorhinus* and to *Calcochloris obtusirostris*, but the relationship is immaterial, an academic construct that shines no light beneath the soils of Africa. In nature, horizontal relationships between similar but disconnected species are of small importance compared to the vertical relationships of disparate species that share the same territory – the interdependent creatures, from invertebrates to carnivores, that keep an ecosystem in balance. That truly is why *Calcochloris tytonis* matters. I had been thinking of it symbolically, as if its value were totemic, its tiny phial like the Tomb of the Unknown Mammal, dedicated to the memory of all lost and dying species. And of course it *is* that. But it's more. The mole is a tiny part of an incomprehensibly complex, infinitely mysterious mechanism that will not work as well without it. I am reminded of something the professor said yesterday: 'Species exist only in our mind. Any sort of living being depends on other living beings, so the evolving unit is never one species. It is always complex.'

Paolo Agnelli, who seems to understand the workings of my

camera better than I do, takes a photograph of the Petri dish, then one of me with the professor. I take one of the two scientists standing together in the curator's office. I have one last question. The Giohar Irregulars failed to find any living examples of the mole in 1964, and no one has recorded any since. Does the professor believe they could still be there, undisturbed by the human turmoil in the world above, silently going about their business? He answers without doubt or hesitation. 'Yes. Why wouldn't they be?' If they were there before the owl pellet, then why shouldn't they be there afterwards? The Petri dish might contain all the known evidence for the mole's existence, but that shouldn't be confused with the species itself. I think again what I have been made to think so often during this brief ascent from ignorance. There is more to the world than the eye can see, or than the imagination can embrace. All we can do is go on looking and listing. The book has been a quest; so has our journey through the museum. From the very bedrock of the planet we have travelled through every kind of articulate life: insects, reptiles, fish, birds, mammals, and onwards deep into our own corporeal entities. Every scrap of it in its own way is a miracle that confounds the laws of chance. Every scrap of it should be clung to with a tenacity that confounds the self-absorption of our greedy and destructive selves. Humans have the power of gods; now they need the wisdom to go with it.

It is late morning, and the professor is glancing at his watch. We invite him to join us for lunch, to share a toast to the man and his mammal, but he declines. His housekeeper will be waiting, he says. Slowly, for he is tired now, we walk with him to the bus stop, while he talks about the novels he plans to write. Looking back after we have said goodbye, I see him raise his stick to the approaching bus, and I wish there could be some new species to be named after him: something of high intelligence, grace and stamina, able to thrive in a range of habitats.

Anything but a mole.

Afterword

Researching a book is like pond-dipping in a river. You can dip and sift all you like, but you can't keep up with the flow. Facts, figures, names and dates churn past in unrecordable quantities, and take no account of your last full-stop. Hence the popularity, among authors at least, of the 'afterword' – a few last, frenzied thoughts dashed down before going into print.

In my own case, the remorselessness of flowing water makes an exact metaphor. The bad news floods down day after day in an undammable surge of despair. Numbers of elephants killed. Numbers of rhinos. Numbers of park rangers. Numbers of species shuffling towards the brink. To get the measure of it, here is the tally of seizures at just one port, Hong Kong, in the first eight months of 2013: in January, $1.4m-worth of Kenyan elephant ivory; in July, another 2.2 tonnes, the biggest haul there for a decade; in August, 1,120 tusks, 13 rhino horns and five leopard skins, all hidden in a single consignment of Nigerian timber. One port, three snapshots of a dolorous torrent. Great apes were also being swept away. The UN Environment Programme estimated that some 3,000 were being killed or captured illegally every year.

In Europe there was some better news. The value of targeted conservation was brilliantly illuminated in September 2013 when a coalition of conservation groups, including the Zoological Society of London, published their report *Wildlife Comeback in Europe*. This showed that a number of once hard-pressed species, including the Eurasian elk, grey wolf, Alpine and Iberian ibex, southern chamois and golden jackal, had made strong recoveries, and that the European bison, once extinct in the wild, had been successfully reintroduced in Poland and Belarus.

For lovers of zoological oddities there were moments of sheer delight. An orange-haired member of an obscure family of raccoon-like mammals called olingos turned out to be an even more obscure 'new' species, the olinguito, native to the cloud forests of the Ecuadorian Andes. Delightfully, it resembles a teddy-bear mated with a domestic cat. Even more delightful for me was the discovery at the Museo Civico di Zoologia in Rome of the preserved skin of a previously undescribed species of mole rat. Its origins were uncertain but it was most likely collected at Mogadishu in 1915 (the year it arrived at the museum) or earlier. In glorious addition to the Somali golden mole, therefore, we have the equally rare and even more mysterious Somali mole rat.

Encouraged by the mole, this book has been written in the cause of the small, the obscure and the humble. If you have stayed with me this far, then I hope you might be tempted to go just a little bit further. Good work costs money. Governments, corporations, charitable foundations and individuals are all respectfully invited to dig as deep as they can. Visit the EDGE website, www.edgeofexistence.org, to see how far a pound or a dollar can go. As I write, £10 ($16.16) is enough to keep a camera trap running for a night – the ideal way to survey shy nocturnal animals. A thousand pounds will train a conservation leader.

It could serve as the motto for the furry underworld itself. Every little helps.

List of Illustrations

Unless otherwise stated, all pictures are the author's own.

Further Reading

The following titles were among the works consulted during the preparation of this book. A few of the earlier ones are available in later editions or in facsimile, but some will be found only in specialist libraries.

Baratay, Eric, and Hardouin-Fugier, Elisabeth: *Zoo: a History of Zoological Gardens in the West* (2002)

Barrington-Johnson, J.: *The Zoo: the Story of London Zoo* (2005)

Bartlett, Abraham Dee: *Bartlett's Life among Wild Beasts at the Zoo* (1900)

Buck, Frank, and Anthony, Edward: *Bring 'em Back Alive* (1930); *Wild Cargo* (1932)

Buckland, Frank: *Curiosities of Natural History* (1891)

Fitter, Richard, and Scott, Peter (illust): *The Penitent Butchers – 75 years of Wildlife Conservation* (1978)

Flannery, Tim, and Schouten, Peter: *A Gap in Nature: Discovering the World's Extinct Animals* (2001)

Gordon-Cumming, Roualeyn: *Five Years of a Hunter's Life in the Far Interior of South Africa* (1850)

Hagenbeck, Carl: *Beasts and Men* (1909)

Hinton, M. A. C: *A Guide to Rats and Mice as Enemies of Mankind* (1918)

Holdgate, Martin: *The Green Web – a Union for World Conservation* (1999)

Huxley, Julian: *UNESCO: Its Purpose and Philosophy* (1947)

Marsh, George Perkins: *Man and Nature; or Physical Geography as Modified by Human Action* (1864)

McCormick, John: *The Global Environmental Movement: Reclaiming Paradise* (1989)

Osborn, Fairfield: *Our Plundered Planet* (1948)

Quicke, D. L. J.: *Principles and Techniques of Contemporary Taxonomy* (1993)

Rice, A. L.: *Voyages of Discovery: three centuries of natural history exploration* (1999)

Schwarzenbach, Alexis: *Saving the World's Wildlife: WWF – the First 50 Years* (2002)

Simonetta, A. M.: 'A new golden mole from Somalia with an appendix on the taxonomy of the family Chrysochloridae' *Monitore Zoologico italiano* (1968)

Stearn, W. T.: *The Natural History Museum at South Kensington: a history of the Museum,* 1753-1980 (1998)

Thackray, John, and Press, Bob: *The Natural History Museum: Nature's Treasurehouse* (2001)

Turvey, Samuel: *Witness to Extinction: how we failed to save the Yangtze River Dolphin* (2008)

Wallace, Alfred Russel: *Tropical Nature and Other Essays* (1878)

Wilson, D. E. & Reeder, D. M. (Eds): *Mammal Species of the World: a taxonomic and geographic reference* (2005)

Acknowledgements

Wherever possible throughout the text I have acknowledged those whose wisdom, knowledge or example has helped me to write this book. But there are one or two who need special mention, not all of whom will be aware of how much I relied on them. Chief among these is the team at the University of Queensland, led by Diana O. Fisher, whose paper on the survival of supposedly extinct species first piqued my interest. The idea was fanned into flame by Craig Hilton-Taylor, head of the IUCN species programme at Cambridge, and Paula Jenkins, curator of mammals, who showed me type-specimens of Lazarus species at the Natural History Museum. I need also to remember some old colleagues – Philip Clarke and Brian Jackman in particular – whose enthusiasm for the physical world rubbed off on me over the years, and editors (Robin Morgan, Sarah Baxter) who enabled me to broaden my experience.

The visit to Ol Pejeta could not have been accomplished without Mark Rose, Ally Catterick and Richard Lamprey at Fauna & Flora International, Richard Vigne and the staff at Ol Pejeta, Andy and Sonja Webb at Kicheche Camp, and my guide Andrew Odhiambo. I am indebted also to Rainbow Tours, who expertly dealt with the logistics and generously picked up the bill.

Without help from the staff at the University of Florence and La Specola, I would never have tracked down the two real heroes of my story, the extraordinary Professor Simonetta and the vanishingly rare creature that gives the book its title. The entire book stands as acknowledgement to the professor himself.

My understanding of the zoological small-print was greatly helped by the London Zoological Society's EDGE project, and I thank its programme director, Jonathan Baillie, for helping me crystallise my thoughts.

The idea for this book – to locate a minuscule bone fragment found inside an owl pellet – cannot have been the most enticing proposal a publisher has ever received. For this reason I am more than usually grateful to my agent Karolina Sutton for selling the idea, and to Poppy Hampson at Chatto for actually buying it. Text editors – the painstaking individuals who go through a text line by line, combing out the fleas – are the unsung heroes of the publishing trade, and in Alison Tulett I had a classically hawk-eyed specimen of the breed. My thanks to her, and to my patient friend Oliver Riviere for his technical help with the pictures.

My wife Caroline was there at the beginning – without her, the idea would have been stillborn – there as a guiding hand through the writing, and there at the end when, in Florence, the golden mole finally became a golden moment. What I owe her is far beyond the scope of the printed page.

Index

Printed in the United States
by Baker & Taylor Publisher Services